UNRAVELING
PILTDOWN

RANDOM HOUSE  **NEW YORK**

UNRAVELING
❧ PILTDOWN ❧

The Science Fraud of the
Century and Its Solution

JOHN EVANGELIST WALSH

Grateful acknowledgment is made to the following for permission to
reprint previously published material:

AMERICAN ANTHROPOLOGICAL ASSOCIATION: Excerpt from "The
Piltdown Hoax" by S. L. Washburn (*American Anthropologist*, 55:5,
pt. 1, December 1952). Reproduced by permission of the American
Anthropological Association. Not for further reproduction.

MACMILLAN MAGAZINES LIMITED: Excerpt from "New Light on the
Piltdown Hoax" by L. Halstead (*Nature*, v. 276, 1978). Copyright
© 1978 by Macmillan Magazines Limited. Reprinted by permission.

NATURAL HISTORY: Excerpt from "The Piltdown Conspiracy" by
Stephen Jay Gould (NH, 1980; vol. 89). Copyright © 1980 by the
American Museum of Natural History. Reprinted by permission of
Natural History.

OXFORD UNIVERSITY PRESS: Excerpts from *Piltdown: A Scientific Forgery*
by F. Spencer (Oxford University Press, 1990). Reprinted by
permission of Oxford University Press.

Library of Congress Cataloging-in-Publication Data
Walsh, John Evangelist
Unraveling Piltdown: the science fraud of the century and its solution /
John Evangelist Walsh.
p.   cm.
Includes bibliographical references and index.
ISBN 0-679-44444-0
1. Piltdown forgery—History.   I. Title.
GN282.5.W35   1996   573.3—dc20   95-9399

Random House website address: http://www.randomhouse.com/

Printed in the United States of America on acid-free paper
2 4 6 8 9 7 5 3
First Edition

*Dedicated
as a welcoming salute to
Eric William Marriott
who came aboard
6 March 1994*

That the bones of Theseus should be seen
again in Athens was not beyond conjecture,
and hopeful expectation; but that these should
arise so opportunely . . . was an hit of fate and
honour beyond prediction.

SIR THOMAS BROWNE, *Urn Burial*

Acknowledgments

A number of very obliging people have made my task in research with this book much easier than it might have been, as well as more pleasant. To each of these I offer my sincerest thanks:

At the library of the Sussex Archaeological Society, Barbican House, Lewes: Joyce Crow, chief librarian; also librarians Susan Bain, Brigid Giles, and Isabel Secretans: the impressive resources and accommodating spirit of the SAC library were of material assistance;

at the University of Wisconsin—Memorial Library and the Geological Library—the unrivaled resources and service contributed significantly to the prompt completion of my work;

at the Natural History Museum, London: Dr. Ann Lum, paleontology librarian, and Dr. Christopher Stringer, Human Origins Group, Department of Palaeontology; also, Robert Kruszynski; the Trustees of the museum for permission to quote from the Piltdown papers;

at the Royal College of Surgeons, London: Dr. Ian Lyle, librarian; at the Marylebone Library, London: Catherine Cook, librarian, The Sherlock Holmes Collection;

Dr. Frank Spencer, Department of Anthropology, Queens College, New York City, for various referrals (and whose careful edition of *The Piltdown Papers* has made the bulk of the basic Piltdown documents handily available to all);

Mrs. Margaret Hodgson (since deceased) and her daughter, Mrs. Ruth Niblett, both of Canterbury, Kent;

Mr. Malcolm Payne, curator, The Conan Doyle Room, Cross Hotel, Crowborough; Mr. Christopher Roden, editor, *ACD, The Journal of the Arthur Conan Doyle Society;*

Mr. Norman Edwards of Uckfield, area historian; Mr. Ian Maitland, present owner of Castle Lodge, Lewes; Mr. John Lucas, Castle Precincts, Lewes; Mr. Mark Lambert, present owner of Barkham Manor, Piltdown; Mr. Geoffrey Denton of the firm of Dawson/Hart, Uckfield; the efficient staffs of public library branches in Uckfield, Lewes, Brighton, Hastings, and Crowborough; also The British Library, London.

For special assistance of an indispensable sort I am indebted to Matthew O. Walsh. For insightful readings of the manuscript, I thank Timothy A. Walsh, University of Wisconsin, William J. Marriott, Wayland, Mass., and Prof. Victor L. Hilts, History of Science Dept., University of Wisconsin; also my patient wife Dorothy for various kinds of essential help.

Any nonfiction writer who deals with an intricate and long-disputed topic must owe a large debt to his predecessors, for information, for stimulus, and for helping to clear the ground. My own debt to the many able Piltdown investigators who have been before me in the field, living and dead, scholarly and journalistic, may be read in the Bibliography and the Notes.

Contents

Prologue: Unfinished

Buried directly beneath the carpet of green grass on which I stood lay the remnants of the bed of ancient gravel that, eighty years ago, so dramatically yielded the bones of Piltdown Man. Then the stretch of grass was an open pit, twenty yards long by some ten feet wide and five deep. At the finish of digging in 1950 it had been kept on display as a memorial, but for many years now it had been filled in. Close by, a plain stone pillar informs visitors, "Here in the old river gravel Mr. Charles Dawson, F.S.A., found the fossil skull of Piltdown, 1912–1913." In England some things tend to change slowly, so the fact that the inscription no longer tells the truth bothers no one.

From where I stood near the memorial stone, there stretched around me the extensive grounds of ancient Barkham Manor, in Sussex, just west of Uckfield. Formerly a large farm, it is now changed into a successful winery, rare in England, with row on row of grapevines marching across what had once been uncluttered fields. The winery's brochures make much of the Piltdown link, proudly giving the name to one of its brands.

Except for several smallish trees, now missing but plainly visible in old photographs of the pit, all else in the area appears as it was then. There is the big old manor house farther along, the gardener's cottage on the same side down a way, the border of thick evergreen hedge running beside the pit site, the long entrance road still lined by a graceful phalanx of thin trees, tall and feathery. From where the trees halt, the road continues, leading past the pit then branching into the manor house and around to the right in the direction of several large outbuildings.

It is a quiet spot, wonderfully serene, with something of elegance despite the grapevines. Walking at leisure here on a sunny day it is easy to forget that in these pleasant surroundings began the greatest, most far-reaching scientific imposture of the last few hundred years (some would say of all time, not without reason), a fraud whose author and method have yet to be unmasked.

The Piltdown hoax—though that is much too mild a word for an event never intended as a lighthearted prank—during the four decades that passed before it was exposed, played a pivotal role in one of the most critical scientific pursuits of modern times, the theory of human evolution. Appearing on the scene just as the fossil record of man had slowly begun to accumulate, but well before any firm answers had been found, Piltdown seriously delayed and skewed the urgent work of science in its search for the truth of man's descent. It created, as one scientist recently expressed it, what was easily "the most troubled chapter in human paleontology," with the fraudulent bones receiving "nearly as much attention as all the legitimate specimens in the fossil record put together." Young scientists and old alike wasted untold thousands of hours on the Piltdown phenomenon. The laborious study, and the writing and publishing of the several hundred research reports and papers worldwide, the sheer, enormous amount of space in books and articles given to sober discussion of its every smallest aspect, make a picture sad to contemplate.

Here is no cause for derisive accusation, as happens too often, no occasion for snide humor. The Piltdown fraud was nothing short of despicable, an ugly trick played by a warped and unscrupulous mind on unsuspecting scholars. All the men involved, those conscientious scientists who first eagerly welcomed Piltdown Man, acted openly and in good faith. None had the slightest reason to doubt the plain evidence of his eyes. A clever antagonist (in this case it was a brilliant one), working underhandedly as he preys on trusting colleagues, can almost always score an initial triumph, even with victims otherwise astute. How long such an imposture may live thereafter depends less on the skill of the forger, or the gullibility of the target, than on circumstances beyond the immediate control of either. In the matter of Piltdown, on three or four occasions at its start the developing affair threatened to collapse because of the forger's astonishing boldness. But each time it was rescued by the fortuitous favorable circumstance.

My purpose in the following pages is not to supply an exhaustive history of the Piltdown affair, with its full geological and paleoanthropological background, including tangents, blind alleys, and incidental occurrences. That demanding task has already been finely accomplished in several competent books and articles by a variety of qualified experts. Instead, I plan, first, to do what has not yet been adequately done: tell the essential story of the discoveries themselves, and the subsequent exposure, just as it all happened, in a continuous, detailed narrative. I shall highlight the personalities involved and, where necessary for full comprehension, describe vital aspects of the science. In this manner I hope to prepare the way for my own solution as to the perpetrator, with guilt demonstrated solely on the evidence.

Missing from all previous tellings of the Piltdown story, even the original newspaper reports, is precisely this factor of a reliable narrative in detail of the actual finds, or supposed finds, as they occurred in sequence. Until a full description of the operations at the pit is to be had, minutely answering the fundamental questions of *what* happened, exactly *when* and *how*, the large amount of hard evidence now on hand cannot well be brought to bear. My first concern, then, has been to provide such a complete narrative, drawing on all available sources, published and unpublished. To that end, where called for I have not hesitated to quote extracts verbatim from the original documents. These give the indispensable, firsthand testimony of participants and observers, and may be looked on as sworn affidavits or depositions. Only one, a crucial one, is of some length.

The use of a narrative framework also permits the varied activities of all those now branded as suspects to be demonstrated rather than simply explained, as has usually been the case. The role played by each of the participants, I believe, is more readily and firmly grasped when their comings and goings are presented in terms of a consecutive, of course strictly factual account. In that regard, I may add that in the first four chapters—describing the discoveries and what was made of them by the world of science—will be found a description of the part taken in the developing story by each of the principal suspects. Questions of guilt or innocence are not addressed in this opening section, however, but are deferred to later chapters.

This much also I would like to have plainly understood regarding the use of narrative: no slightest fact or reference is an invention, nothing has been imagined. All is derived from one or another of the manifold sources, or comes from my own observation on the spot. If I say that someone went somewhere or did something, or responded "coldly" or spoke "heatedly," it is because the documents say or clearly imply so, or the known and living situation requires it. (The extensive Notes section, of course, allows a full check on everything claimed or displayed in the narrative.)

Crucial to any resolution of this extraordinary affair is a full exposition, step by specific step, of *how* the deception was actually done, in this instance a factor inseparable from the question of *who* did it. Vague references to the bones having been "cleverly broken," or to the pit having been "ingeniously salted," of course, do not come near answering the questions raised. The Piltdown forger was as diabolically clever in his planting of the doctored specimens as he was in secretly preparing them. That is a point, perhaps the main one, never fully understood or appreciated, apparently not even by the men who exposed the hoax. But in the end it proves pivotal, as I believe my final chapter shows.

The original Piltdown bones—only the skull was found, including part of the jawbone, in fragments—while now thoroughly discredited, have not been lost. They are carefully preserved in London's Natural History Museum, where initially in 1912 they were brought for study. My own first sight of them took place there in October 1993, in a room of the Paleontology Department, Fossil Hominids. Very kindly, I was allowed to examine and also photograph them, making suddenly alive and real all that I had read during many years about the Piltdown story.

In color they are rather unexpected, being an intense, almost vivid shade of deep chocolate brown. All are now covered by a glossy, transparent coating of preservative, so that their appearance is not quite what it was when, at different times, they were lifted out of the dark brown gravel at the Barkham Manor pit.

Each piece of bone in turn I took up and handled, the broken half-jaw and the several sections of the shattered cranium. Curiously I ran a finger over the smooth surfaces, tracing the delicate bulges and curves, and the rough edges where the breaks had occurred. It was a fascinating, nearly hypnotic experience, particularly in noting the extraordinary thickness of the cranium wall, twice that of a normal skull. The two molar teeth, still fast in their sockets in the simian jaw, were indeed remarkably flat, showing what has been identified as a pattern of human wear.

In my right hand I held the jawbone. With my left I picked up the largest piece of the cranium, a dark, slightly concave fragment some four inches across. One American paleontologist, I recalled, the Smithsonian's Alex Hrdlicka, long before exposure of the fraud in 1953, had done the same thing in the same museum, and had reported an unexpected result: the jawbone was lighter in weight for its mass than the piece of cranium. Hrdlicka had failed to follow up on his unanticipated insight, but it had real significance, for the jaw, as is now known, is not nearly so fossilized as the other fragments of the skull. For myself, gently hefting the two pieces, there was only confusion. At one moment I decided they weighed about the same. The next I felt sure they were slightly different.

Even armed with all the great knowledge possessed by those men who first dealt with these bones, even with the knowledge acquired since, I would not have wanted to be the one responsible for making a judgment on them. The men who did in fact render that original judgment might easily have hung back in the safety of temporizing and equivocation. But with all the courage of their convictions, moved by an intense desire for knowledge in a new and promising field of study, they boldly declared themselves. Though proved wrong, caught napping over their own expertise, they are entitled to more respect, I would insist, even some admiration, than they have yet found.

Too often in magazine articles and books in which attempts are made to solve this unique mystery, the discussion of possible culprits, of motives and methods, proves disappointingly superficial. Frequently, in place of solid evidence or graphic demonstration, there is offered only bland supposition and glib suggestiveness, turning what should be a serious analysis of meaningful events into something like a parlor game. As a result, accusations of guilt have been lodged by different investigators against more than a dozen men who, to one degree or another, were involved in Piltdown, some of the connections tenuous in the extreme. This, to me, of all Piltdown's results is the most to be regretted. Years after their deaths, some of science's worthiest names—and one literary name, worthy indeed—have been blithely charged with having lived a lie, led double lives, almost schizophrenic.

Cogent reasons of intellect and scholarship have been suggested for pursuing, even at this late date, a resolution of the case, for baring the forger's identity and revealing his methods. Compelling is one reason in particular: the rare chance afforded to learn more about the interior process of science, in itself and in its relation to society. Another reason cites the opportunity to see and understand scientists themselves in more ordinary guise as less than perfect people.

Especially relevant is a third reason, uncovering the pernicious effect on science of fashionable ideas, "the degree to which a prevailing paradigm may influence and even dominate not only thinking but discovery." That particular danger is greatest, of course, in pioneering studies where the knowledge required is less than adequate, making the hunt for answers as much art as science, as with Piltdown.

Such reasons are commendable, certainly, and investigators are right to invoke them, as they invariably do, in justifying the large amounts of time even now being expended on a problem that many shrug off as beyond resolution. Yet none of these reasons, I feel, can be as imperative as the simple human duty, incumbent on all, of upholding in death an untrammeled record of the good work any of us may have done in life.

Here, assuredly, is the most pressing reason why the Piltdown puzzle, all these long decades afterward, still demands to be solved. It is a matter of protecting the innocent. The Piltdown forger must finally be brought to book so that the other suspects may go free.

UNRAVELING
PILTDOWN

1

Curtain Falling

In comfortable retirement in his tidy Sussex cottage, Arthur Smith Woodward found his days scarcely less occupied than they had been during his forty crowded years at the Natural History Museum in London. Still youthful at sixty, his reddish hair and trim Vandyke beard as yet only touched with gray, he was to remain active in paleontology for almost another two decades. It was an unusually happy retirement, shared with his wife and, until her marriage, a daughter, providing a rarely satisfying close for one of the most respected and consequential scientific careers of the time. The ultimate honor came just as he took leave of his duties, in the spring of 1924: conferral of a knighthood.

Woodward's eminence had been fairly won. With no advantage of wealth or background, lacking even a university degree, by sheer brilliance and unflagging energy he had worked his way upward at the museum, at thirty-seven becoming keeper (director) of its world-famed geology department. Along the way his many outstanding achievements in paleontology (mammalian, reptilian, and fossil fish) had been fully recognized. A fellow of the Royal Society before he was forty, by the time of his retirement he had been president at different times of three prominent scientific bodies. A recipient of the Gold Medal of England's Royal Society, he had also been awarded the Lyell Medal, the

Linnean Medal, the Wollaston Prize, the Prix Cuvier of the French Academy, and the Thompson Medal of the American Museum. Aside from his half-dozen technical books, the total of his scientific writings exceeded a remarkable four hundred papers.

Capping all was the association that had brought him fame and secured his place at the pinnacle of his profession, the central role he took in the drama of Piltdown Man. Hailed by most as evolution's first true "missing link," that sensational find since its arresting debut in 1912 had usurped a large portion of his time at the museum, and through his twenty-year retirement it continued to rivet his attention. Attacked, defended, and argued over, its meaning and its very nature were continually being reevaluated and reinterpreted by the world's scientists. To a reporter from a London paper who interviewed him on the day he quit the museum for good, Woodward readily admitted that the Piltdown discovery had been "the most important thing that ever happened in my life."

Retirement had also brought the deserving Sir Arthur an unexpected dividend, a welcome softening of attitude and outlook. In his professional capacity a man of great reserve, he had been fiercely dedicated to his work: except when traveling on business for the museum, during forty years he had missed no more than a day or two at his desk. Perceived as cold in temperament and baldly impersonal in style, by colleagues no less than subordinates, he had fed that perception by standing decidedly aloof. Now, freed from the constraints of office, his rigid manner relaxed into that of the gentle-tempered man his family had always known. At ease with himself, more casual in deportment, he spent much time with children, happily taking them on excursions to zoos and museums. His daughter, Margaret, who as a child had frequently accompanied him to the original excavations at Piltdown, later chose to describe him in retirement with a word that would never have occurred to his peers: "jovial."

From earliest youth, it seemed, nature had formed Woodward for a scientist. At six or so, just as he began school, he became fascinated by a pig's head in a butcher-shop window. He persuaded his mother to buy it, and then had her slice open the ear canal so he could see where it led and how the ear worked. "I was deeply disappointed," he soberly recalled in his old age, some seventy years later, "when I found that the passage ended blindly." He also induced his long-suffering mother to skin and boil a dead mole so that he might strip away the flesh and have a clean skeleton to study.

His first scientific paper, though never published, was written while he was some months short of his fourteenth birthday, an account of some shells he picked up at the seashore during a family holiday in Wales. As an old man he would still proudly insist that the youthful composition had been a creditable performance "without amateurish faults." He could also remember just when and how his decision to become a geologist was sealed on him: it happened during another holiday at the shore when he met an old man, a former copper miner, who gave him a fossil shell of wondrous shape, "which he told me he had found deep down in a mountain."

Prophetically, in a way, Woodward's most vivid schoolboy memory concerned a huge model of a molar tooth he often passed in his home town of Macclesfield. Suspended over the entrance to a local dentist's office, the monstrous tooth hung well out over the pavement, and he would stop and stroll back and forth beneath it, head thrown back as he gazed up in boyish wonderment. With Piltdown, decades later, a high proportion of his mature effort was expended in close study of two curious molars.

At eighteen he enrolled in Owens College, Manchester. But after two years, when he placed first in a competitive examination for a rare and coveted opening at the Natural History Museum in London, he took his departure. Not only had he been the youngest candidate, he was also the only applicant without university credentials. Barely a decade at the museum brought him wide recognition as a leading authority on nonhuman fossils, especially of ancient fish, a vast and fairly new field in which it was soon acknowledged that he had few equals.

By no means, however, did Woodward's success come without heroic effort, even some anguish of spirit. His lack of formal schooling was a constant concern, and he did what he could to repair the deficiency, for years attending evening classes at King's College in London. He had almost no personal life. On first reaching the city he managed for lodgings only the poorest of accommodations, during two interminable years inhabiting a single tiny bedroom in a nondescript neighborhood. Outside the office he had no acquaintances, so that for months on end, as he recalled, he "spoke to no one from the time he left the museum each day until he returned the next morning." It was a lonely and depressing existence for an earnest and ambitious young man, and inevitably he lost himself in his work. Geology, fossils in particular, came to occupy nearly every moment of his waking thoughts, so it was only fitting that he should have acquired his wife through the same channel. In 1894 he married Maude Seeley, the daughter of a professor at King's College, the well-known geologist Harry Seeley, whose classes he attended.

During his retirement, from the mid-twenties onward, Sir Arthur kept abreast of every twist and turn in the fast-developing evolutionary picture, being deeply involved in the discussions on all the more significant fossil discoveries made throughout the world. In the evaluation of the four major hominid finds—*Australopithecus* and *Rhodesiensis* in Africa, *Sinanthropus* in China, and *Swanscombe* in England—as well as the ever-expanding record of the intriguing Neanderthals, his voice was prominent. Each of these finds, in turn, had to be measured against Piltdown, driving the investigation of Piltdown itself along fresh and unexpected paths, and bringing it under ever more searching scrutiny.

The Piltdown mandible (jaw), especially, precipitated loud disagreement, as it had from the first. A jaw so thoroughly apelike, critics insisted, simply did not belong with a cranium (brain case) so undeniably human. The jaw's manlike features, all of them delicate and disputable, hardly entitled it to be taken as part of a human or even half-human face. Piltdown, it was charged, had mistakenly been manufactured from two separate creatures, a fossil man and a fossil ape: the remains of the two had just happened to come together in the ground, a freakish prank of nature. Combining them only created a monstrosity that had never in fact existed.

Led by Woodward, Piltdown's advocates countered that the two fragments had been found buried in close association, making astronomical the odds against their being unrelated. Also, the distinctive wear pattern plainly to be seen in the two molar teeth, their pronounced flatness, could be attributed to no one but a human. True, the Piltdown combination of animal jaw with human cranium strangely and precisely *reversed* the evolutionary pattern steadily emerging as newer fossil skulls came to light. Still, it remained legitimate, insisted Woodward and his colleagues, a necessary step in the rise of modern man. Not a few paleontologists, in any case, firmly believed that, as Piltdown indicated and despite the accumulating evidence to the contrary, in man's descent the brain *had* led the way. For them, as for Woodward emphatically, an increase in the size and complexity of the brain came first, preceding all bodily adaptation such as an upright posture or a decrease in muzzle size. With its advanced brain and beast's face, Piltdown seemed a clear proof of the theory.

Woodward's continuing intimate link with the Piltdown fossil had even dictated his choice of a retirement location. His house in Sussex—called Hill Place and situated on the western edge of the busy old town of Hayward's Heath—was within walking distance, at most a short ride, of Barkham Manor. For a dozen or more years in his retirement, during every spring and summer but one, he spent many days and weeks digging earnestly in the old gravel pit, supervising the movement of large deposits of soil, having every shovelful carefully sifted, himself often taking a hand with the sieve. The fact that week after week and year after year he encountered nothing further, no additional bones or flints or other materials with which to expand the Piltdown portrait, failed to subdue his dogged search. It was advancing age, on top of physical disability, that brought a halt to his annual visits to the pit, and only late in the thirties did he finally concede an end to his search there. Then he arranged for a five-foot memorial stone to be erected beside the spot, now filled in and grassed over.

Dedication of the stone in July 1938 was an occasion of some consequence, attracting dignitaries from the surrounding towns and London. The main address was given by Sir Arthur Keith, England's leading anatomist, and an old foe of Woodward's in the heady days of the original Piltdown controversy. The two men had wrestled mightily over the size of the Piltdown brain, among other things, but in old age had become fast friends, more in agreement than they had ever thought possible. Woodward and Charles Dawson, the original discoverer, said Keith in his few words of praise, between them had uncovered "a long past world of humanity such as had never been dreamt of."

Only one really sour note intruded on the serenity of Woodward's retirement, an unexpected charge that he had once been made the butt of a small hoax. It had started innocently enough in 1914, when some boys at a school in Dorset reported finding a piece of ancient bone in a quarry. On the bone was scratched in thin but sure lines a rude drawing of a horse's head and forequarters. Sent for inspection to the Natural History Museum, the bone soon reached Woodward's hands, and he readily identified it as a rare and remarkable example of "the pictorial art of Palaeolithic man." There existed one other such specimen, he explained in the journal of the Geological Society, and interestingly enough this new bone-sketch was "almost identical with the first, both in subject and in style." No more was heard of the subject for more than a decade, until the first year of Woodward's retirement. Then another leading paleontologist, William Sollas of Oxford University, included a casual statement about it in a later edition of his book, *Ancient Hunters*. "The bone described by Dr. Woodward," remarked Sollas in passing, "is a forgery perpetrated by some schoolboys."

Indignantly, in an article in *Nature* Woodward repulsed the idea of a counterfeit. He had been assured by the boys themselves, he stoutly declared, that there had been no trickery. Both bone and sketch, he insisted, were genuine. In the same magazine the following month, Sollas repeated his charge, giving specifics and naming his sources. The matter went no further at the time, at least in public, but all too obvious was the fact that the rough sketch on the new bone suspiciously matched that on the undoubted older specimen. It was not an important find, however, and apparently both parties were glad to let the subject drop. (Fifty years later this almost ephemeral incident would be resurrected to play an unexpected role in the Piltdown story.)

The World War Two era brought its own, more personal hardship to the Woodward household, as Sir Arthur faced the unsettling threat of dimming eyesight. Despite treatment, it continued to fade, and by the time he had passed his seventy-sixth birthday he had become totally blind. Still he refused to give up or give in, and when his wife suggested that he resume a project that he had begun and let lapse nearly three decades before, he responded eagerly: the writing of a book on Piltdown.

In 1916, when the impact of the original discoveries was still vivid, he had written an opening chapter of the proposed volume, sketching his own part in the unique drama. Forced by the press of work to lay the manuscript aside, he had somehow never returned to it. Now in his blindness—as the skies over Hayward's Heath filled almost daily with Royal Air Force planes desperately fighting the Luftwaffe in the last-ditch Battle of Britain—Woodward regularly dictated to his patient wife all he had to say about the history and meaning of the Piltdown fossils. In the summer of 1944 he fell ill, but with his old tenacity he continued to dictate from the depths of his bed. Laboriously he spoke the book's final sentence on the first day of September. The following morning, three months short of his eighty-second birthday, he died.

Confidently titled *The Earliest Englishman*, the book was published in 1948. A small volume making just over a hundred pages, it began by neatly if rather too sketchily summing up the story of the original finds, and followed with lengthy treatment of various aspects of the topic, such as soil composition, the age and extent of the gravel beds, and associated animal fossils. He also offered some informed speculation about the Piltdown race. The creature had mastered the use of fire, he thought, ate meat, nuts, and roots, made crude tools of flint and bone, and dressed entirely in rough animal skins. It certainly could talk, "expressing himself at least as well as any of the existing savages." But no longer did Sir Arthur see Piltdown Man as a full-fledged missing link. Now he felt sure it had already joined the human family. "He certainly was a man, and not a creature halfway between man and ape. He was perhaps ungainly, and may have walked with a shuffling gait, but his brain and skull were essentially human, only with a few ape-like traits."

Also acknowledged in the book was the continuing problem of the jaw, touching as it did on a fundamental point of evolutionary mechanics. During the decades since Piltdown, many more fossil skulls had been found. Without exception, all showed not a steep human forehead rising above an ape's chinless, jutting jaw, but just the opposite, a more graceful human jaw supporting a beetle-browed vault. Evolutionary change, it seemed, had shown itself first in the physical body, altering form and function, well ahead of any advance in the brain, and presumably in mental capacity. To this conclusion, inescapable on the evidence, Piltdown was the only exception.

Aware of this opposite trend in the accumulating fossil record, Woodward sensibly adjusted his position. Developing man must have advanced along two or more separate lines, he decided, and of these all but one, ill-fitted for their purpose, had become extinct. Which of the lines had prospered, leading to modern man, was for most scientists still the vital, much-argued question. For Woodward, however, the answer was abundantly clear. Piltdown may have borne many marks of the ancestral ape, but he was "indeed a man of the dawn," a progenitor of the existing human race.

Woodward's small book appeared just as England's scientists were returning with renewed zest to the war-delayed problems of human evolutionary theory. Technical though much of the volume was, its personal tone and wealth of authentic detail helped to spark a lively renewal of interest in the original Piltdown discoveries, and a call soon went up to do something about preserving the site of the excavations. Under government auspices, after some final excavations, the small plot of ground was bricked in, with the precise spot of the discoveries being kept open and protected behind thick glass. Piltdown had become "a major event in the unfolding of man's remote past," it was declared, and the ground that had yielded the fossils would have great historical value "for unborn generations." When in the spring of 1950 the almost forty-year-old site was thrown open for public viewing, it quickly became a focal point for tourists and school outings.

It was in this same year that the first puzzled suspicions, ironically triggered by a wish to obtain the clinching evidence for authenticity, began to stir.

Late in 1949 the bones were taken from the vault of the Natural History Museum and submitted to a test that had been only recently perfected. The new procedure, it was thought, would settle the vexed question of the jaw-cranium association. Fossil bones absorb fluorine from soil water. If the Piltdown bones had lain in the same patch of earth for the same length of time, they should have absorbed the same amount of fluorine, which could be measured. Should the ages of jaw and cranium coincide, the fact would greatly strengthen, though it wouldn't prove, the claim that they belonged together. On the other hand, a clear disparity in ages would almost certainly decide the question in the negative.

All the work of testing was done on neutral ground by a government chemist, "every available bone and tooth from the Piltdown gravel" being analyzed, and the results were ready by year's end. Happily for Piltdown's supporters, the association was fully confirmed: the amount of fluorine in the jaw did indeed match that in the brain case. The two specimens were "of the same age." But the test also gave a second result, based on further arcane computations—a result never envisioned.

If the fluorine tests were accurate, Piltdown was not nearly so old as everyone had believed, but must be assigned to the late Upper Pleistocene. It could be dated, in other words, as recently as fifty thousand years, nothing like the half-million or more years that had become standard. Piltdown Man, it appeared, had lived in an era when the earth already held "many examples of fully developed men of our own time," beings who, while no doubt somewhat brutish in manner, were completely human in body and mind.

Unless there was an egregious error in the test calculations, Piltdown Man was not anywhere near a "dawn man," let alone a missing link. He was a shocking anachronism, an impossible survival out of a dim and far-distant past. To his contemporaries, even primitive as they themselves must have been, he would have appeared as an alien, if not a veritable monster. There was no possibility that he had played any sort of role in man's descent. By the time Piltdown Man came on the scene, humans were already established and at the point of taking control of the environment.

From scientists, this curious and disturbing news brought varying degrees of concern. Not all were alarmed by it. Some felt that the newly devised fluorine test must be flawed, or the measuring apparatus had been too crude or else ineptly applied. Others accepted one or another of the several geological explanations offered, in particular one that drew on the unpredictable whimsies of "genetic isolation." At the time of the great Riss-Würm interglacial period, it was suggested, the island of Britain may already have been severed from the continent by formation of the channel. This physical separation could well have favored the rise of freakish anatomical throwbacks.

Among those who were slower to decide the question was a young professor of anthropology at Oxford University. Fascinated by the idea of a creature part man and part ape inhabiting a world that had passed it by, he continually turned over in his restless mind every smallest aspect of the Piltdown story. His name was Joseph Weiner. It would require several more years, and a seemingly casual conversation at a banquet, before the critical point was reached. But it was Weiner who would, initially in some reluctance, startle the world of science by daring to breathe aloud the word forgery.

2

The Oldest Humanoid

Walking up the broad, sweeping approach to the grand entrance of the Natural History Museum, Arthur Woodward, keeper of the museum's geology department, went through the high-arched doorway into the magnificent main hall. Bidding a guard good morning, he turned right a few steps and pushed open the door to his office. A large, squarish room with high windows, it was full of heavy oak furniture, its walls lined with bookshelves and cabinets. Outside the windows, beyond the tall iron railings that enclosed the museum grounds, there was a view of busy Cromwell Road crowded with a hurrying, two-way jumble of horse-drawn cabs and motor-driven cars and buses. It was Tuesday, February 15, 1912, a date the keeper would have cause to remember.

Already on his desk was the day's first mail, and he began flipping through the envelopes, occasionally opening and reading one. When a cancellation from the Sussex town of Lewes caught his eye he easily recognized the small, assured handwriting of the address. It was from his friend Charles Dawson, a solicitor by profession but an amateur geologist and antiquarian of some standing.

The letter filled three pages of the folded sheet, with the last page left blank. It was a reply to a note Woodward had written Dawson some days before, following a trip the two had made together to a quarry near Hastings to inspect some dinosaur bones recently uncovered. In his note Woodward had asked about paying his own share of the expenses, and had also inquired about a curious rumor he had heard regarding an acquaintance of Dawson's, the writer Arthur Conan Doyle. An unusual novel, Woodward understood, was being written by Doyle, one which would have special interest for geologists. Had Dawson heard anything about it? Supposedly it concerned prehistoric animals.

In his answer, Dawson estimated that two pounds five shillings would cover Woodward's part of the expenses at Hastings, laid out for transportation, for the man who had done the digging, and as a fee to the quarry foreman. As for Doyle's new novel, Dawson wrote that his friend was indeed composing a story to interest geologists: "a sort of Jules Verne book on some wonderful plateau in S. America with a lake, which somehow got isolated from Oolitic times, & contained all the flora and fauna of that period, and was visited by the usual 'professor'."

Doyle and Dawson were fellow members of the Sussex Archaeological Society, and in the previous November Dawson and his wife had been guests in the Doyle home at Crowborough, just north of Lewes. It was a social visit but also a working one, for Dawson had been invited to comment on what appeared to be some dinosaur bones found in the vicinity (he decided they weren't). If while he was at Crowborough he heard the title of Doyle's book (*The Lost World*), he did not supply it in his letter, and nothing more was said at the time about the strange new tale.

But Dawson's letter also carried other news that, for Woodward, would have been far more intriguing than any fictional excitement. He had come upon some very ancient soil deposits, wrote Dawson, quite extensive, which might easily yield interesting fossils. They overlaid the old and well-known Hastings gravel bed at a spot near Uckfield: "It has a lot of iron-stained flints in it, so I suppose it is the oldest known flint gravel" in the area. The letter's next sentence would have thoroughly arrested Woodward's attention, though he must have read it several times before it made good sense. Obviously, in his hurry Dawson had inadvertently dropped a phrase. The sentence read: "I think portion of a human (?) skull which will rival *H. heidelbergensis* in solidity."

No doubt the truncated thought could be read as *I think I have found a portion*. If so, and if Dawson really had dug up a portion of a human skull to match the ponderous Heidelberg Man, the news was indeed exciting.

Scarcely four years before, a massive jaw had been uncovered at Heidelberg in Germany, as large and heavy as an ape's, and with an ape's sloping chin. But it also showed definite human characteristics, especially in its dentition. Unlike those of an ape, the Heidelberg teeth were small and manlike, with no sign of a great projecting canine. In 1912 the Heidelberg jaw was the only fossil evidence generally accepted as clearly attesting to an ancient, animal-like progenitor for the human race (Java Man of 1891 and some problematic English skeletons were poor rivals). Though not without challenge, it had already taken its place as the prime exhibit in the evolutionary argument, by then on the boil for half a century.

Depending on what Dawson meant by a "portion," and if the bone really was human, thought Woodward, Britain might now have its turn in the spotlight, might be on the verge of making a truly significant, even a decisive contribution to the subject. The expression in Dawson's letter was murky, yet it did seem that the piece of bone had been found in those same ancient gravel beds near Uckfield. If so, there existed a good chance that more of the skeleton would reveal itself.

Eager to inspect both bone and gravel bed, Woodward replied promptly, his note apparently dispatched that same day. However, he was delayed longer than he could have expected, first by the demands of his office, which included an extended trip to Germany, then by wet weather making the roads impassable, as Dawson reported, and then by the subsequent flooding of the gravel beds. Dawson too found himself hampered by his professional duties—besides being a busy lawyer with offices in Uckfield, he held several town offices—so that more than three months passed before the two men met to visit the pit site. During this anxious wait Woodward warned Dawson to keep the find secret, especially not showing it to anyone who had knowledge of the subject. There was no need to worry on that score, Dawson had assured him. He would be circumspect, he replied, would "leave all to you," and would keep away from the site itself lest anyone might be observing his movements. "I have decided to wait until you and I can go over by ourselves," he wrote. ". . . it is not far to walk from Uckfield and it will do us good!"

In May, Woodward was given his first look at the discovery itself when Dawson came up to London on business. Visiting his friend in his office at the museum on the afternoon of May 23, he unwrapped a small package on the desk. "How's that for Heidelberg!" he called out happily.

When he saw what Dawson's word "portion" really meant, Woodward's eyes widened. On the desk lay three separate fragments of thick cranium, including one fairly large and long piece of the left parietal (side). Intriguingly, this fragment still held a small slice of the forehead, tapering down to a tiny piece of the brow ridge, just at the corner over the left eye. The thickness of the wall was pronounced, more so than in any normal human skull Woodward had ever seen, in itself a striking proof of primitive origins. Yet that fact, arrestingly, was contradicted by the parietal fragment. The subdued bulge of the brow ridge, together with the forehead's vertical slope, indicated a brain case no different from the modern type. Even at a glance the skull showed marks of both man and ape.

Woodward's immediate hope that more of the skull might still lie nestling in the ground was enhanced by Dawson's quick preliminary account of how the three pieces had come to light. Workmen digging for road-mending materials in the pit at Piltdown, he said, had partially uncovered the more or less intact cranium. But one of them, not recognizing the object, and perhaps before he was quite aware of it, had shattered the skull with a blow, or several blows, from his pickax. In that first instant, it seems, the skull had been mistaken for a coconut, an understandable error since the iron-impregnated soil had colored it a dark, rust chocolate brown.

The fragments of the splintered skull, soon scattered by the steady shoveling, had been ignored by the busy diggers, except for one man who retrieved a single piece and brought it to Dawson. Diligently searching the disturbed soil, then and on several subsequent occasions, Dawson had turned up only the two additional pieces. If the remainder of the skull had not already been hauled out and spread on the area's roads, a good part of it, at least the cranium, must be left in the pit. For now, digging by the road menders had been halted.

Woodward's position at the museum frequently brought him claims from the public regarding out-of-the-way matters and unusual finds. Most of these he was able to pass off readily, their unlikelihood or spurious nature becoming obvious. With Dawson it was different, for the two had known each other for almost thirty years, and Dawson too in a more limited way had made a name in geological circles.

Like Woodward, Dawson's fossil hunting had begun in boyhood, and by 1884, when he was twenty, his collection of dinosaur remains, gathered in Sussex, had aroused the interest of the Natural History Museum, which promptly bought it. Six years later he made an important if relatively minor find, also in Sussex, discovering a mammalian tooth which gave the first evidence for the presence in Europe of a Cretaceous mammal. The tooth was submitted to Woodward at the museum, who confirmed the identification and officially named it, in the usual way, *Plagiaulax dawsoni*. Enrolled among the museum's honorary collectors, Dawson was at the same time elected to a fellowship in the Geological Society. Besides fossil hunting, he also had wide antiquarian and archeological interests, and his several accomplishments in these areas were admitted to be not only impressive but remarkably varied.

While the two men were not quite personal friends—he was often in London, yet Dawson had briefly visited Woodward's home on only one or two occasions—they admired each other and, though of very different temperaments, got on well together. Dawson, with his outgoing, enthusiastic manner, the easy welcome he found in all gatherings, was almost the direct opposite of Woodward, with his subdued ways and measured personality. But the divergence in their natures had been no bar to what each man admitted was a pleasant and productive association. Woodward was later to recall Dawson as

> . . . one of those restless people of inquiring minds, who take a curious interest in everything round them . . . from an old parchment deed to a horn-like growth on a horse's head; from the proverbial live toad in a stone to an escape of inflammable gas from the ground; from fossils and minerals in the Wealden rocks to the tools and other leavings of prehistoric man. Nothing came amiss to his alert observation . . . He was a solicitor by profession, but during his leisure he lived in the world of scholars.

Barkham Manor, near Piltdown Common in the town of Fletching, in the county of East Sussex, is an ancient estate, one of several in the wide, gently rolling countryside west of Uckfield. Lord of the manor in 1912 was George Maryon-Wilson, who had other extensive land and estate holdings in this part of East Sussex. Tenants of the manor at this time were Robert Kenward and his family (wife, two sons, and a daughter). The Kenward roots in the area stretched back five centuries.

Leading into the estate from the road, stretching toward the manor house, is a long, arrow-straight tree-lined driveway. Fifty yards short of the house the parade of tall trees stops, and the grassy borders on either side of the drive widen. In the right-hand border, just beyond the trees, lay the open gravel pit that had given up the bones.

Once, a great while before, this area had been the bed of a swiftly flowing river, the Ouse, which in more recent times had altered its course. Still full and flowing in 1912, in a series of wild twists and turns it poured its way south through the Weald (an extensive tract of land, formerly a forest, sprawling over three counties), breaching the Downs and emptying into the English Channel at Newhaven. Its nearest approach now to Barkham Manor is a sweeping bend at Newick, just over a mile away.

The necessary permissions for digging at the manor, both of owner and tenant, had been arranged by Dawson, an easy assignment because his firm had a long-standing business connection with the manor's lord, handling the estate's business affairs. In requesting the permission, though, he had been careful to tell a small lie, saying nothing about bones but explaining that he was in search of unusual flints. He also arranged for the necessary workman to be on hand for the shoveling, a local laborer named Hargreaves, who bore the curious nickname of Venus. Over the next few years, Hargreaves would almost always be on hand to do the digging, his work directed by Woodward and Dawson.

Every shovelful of earth had to be minutely inspected, an operation that required putting it through a series of fine meshes while slowly fingering the loose soil, a tedious process. Patience and sustained attention were needed to pick out the bits of bone that would be darkly stained by the soil, distorted by clinging earth, and made almost invisible in a mix of pebbles, concreted sand, ironstone, and brown flints of all sizes. In addition to digging in the undisturbed soil they would go carefully through the "spoil heaps" that lay around the pit, dirt left over after the workmen had extracted the flints for road repair.

As it turned out, there was to be a third geologist on hand for this first day's digging, a young Frenchman already known to Woodward for his work on the fossil flora of Sussex. An ordained priest assigned by his order in France to pursue further study at the Jesuit seminary near Hastings, and a budding paleontologist, thirty-year-old Pierre Teilhard de Chardin had enjoyed a loose association with Dawson since the summer of 1909, when he was still a seminarian. One day both had been in search of fossils at the Hastings quarry, had happened to meet, and had found each other congenial. The young priest, eager to make a mark as a scientist, was happy to accept the older man's offer of guidance in matters of English geology. In turn, Teilhard's youthful eyes and energy, and his knowledge of Hastings stone and gravel beds, Dawson had found impressive and helpful.

Teilhard, in fact, was one of the few to whom Dawson, forgetting what he had promised Woodward, had shown the Piltdown bones (that had happened six weeks before, in mid-April, when Teilhard in a letter mentioned seeing pieces of "a very thick, well-preserved human skull," which he said Dawson "excitedly drew" from a box). It was at Dawson's invitation that Teilhard showed up at the pit on Saturday, June 2, happy to take part in what promised to be a stimulating venture.

Success rewarded the party on the very first day of digging. Not spectacular finds, they were enough to delight the staid Woodward, who welcomed this immediate proof that the pit held further fossil treasures. He reacted, as Teilhard recalled, "with the enthusiasm of a youth, and all the fire that his apparent coldness covered came out." The most important of the finds, occurring after several hours' searching, was another small chunk of bone, a thick piece from the back of the skull, the occiput. It was spotted by Dawson himself, lying in the darkened debris of one of the spread-out spoil heaps. Found by Teilhard a few minutes afterward was a valuable animal fossil, a broken tooth of *Stegodon*, a small dinosaur.

For the next two weekends Dawson and Woodward returned to the excavations, Teilhard being absent because of clerical duties. If Teilhard went again to the pit during that first year, he was present only a small part of the time, his freedom curtailed by his training program at the seminary. Then, in mid-July, his schedule took him back to France, where he remained, not again visiting England until the following year. The exact days and hours of his presence at the pit in 1912 thus are not wholly certain.

After the first finds, the long hours of close inspection turned up only a few worked flints and some bits of ancient animal teeth, interesting enough in themselves and which would prove of great value in dating the human remains. Not until the fourth Saturday of June, with energies flagging and everyone's initial eagerness a little dulled, did there come the sensational find that made all the effort worthwhile: with dramatic suddenness the broken jaw was uncovered.

It was a sultry evening toward the close of what had been a very hot day of digging. All the members of the party were covered with perspiration and dirt from moving about in the loose soil of the uneven trench bottom, and again it was Dawson himself who made the find. Disappointingly, his description of how it happened is short, unspecific and matter-of-fact: "It was not until we had been busy off and on for some weeks that after a hard and unproductive day's work I struck part of the lowest stratum of the gravel with my pick, and out flew a portion of the lower jaw from the iron-bound gravel."

Woodward's memory of the same climactic episode was given in somewhat more detail but curiously is nearly as subdued as that of his friend. After a long, sweaty day of searching, he recalled, Dawson on his own

> . . . was exploring some untouched remnants of the original gravel at the bottom of the pit, when we both saw half of the human lower jaw fly out in front of the pick-shaped end of the hammer which he was using. Thus was recovered the most remarkable portion of the fossil which we were collecting. It had evidently been missed by the workmen because the little patch of gravel in which it occurred was covered with water at that time of year when they reached it.

The piece of jawbone had been dislodged from its ancient bed in the hard bottom soil of the pit by a sharp blow from the geological hammer in Dawson's hand. Until the moment it was freed, it must have been completely hidden from view, for if it had been at all exposed Dawson would not have struck with his hammer at the packed earth. Instead, with a dull-edged tool, brushing as he went, he would have probed slowly and carefully in order to work the prize loose without damage. But he delivered the blow blindly, apparently with some force.

Both Dawson and Woodward picture the jaw's sudden appearance up or out of the dirt as an abrupt jump, taking both, it seems, a little by surprise. For one, it actually "flew" out, implying something more than a slight movement. Similarly, the other saw it "fly out," describing an impetus that would have sent it some inches or feet along the trench's bottom.

Throughout the remaining weekends of the summer, digging went on, but nothing turned up to equal the jaw. One discovery, however, made by Woodward, later proved to be indispensable in the work of reconstructing the skull. Found within a yard of where the jaw had come to light was another small portion of the occiput (back), which exactly fitted the jagged edge of the occipital fragment found on the first day of digging. Joined, the two pieces bridged the gap to the left parietal, helping to settle the question of the skull's size and shape. The new piece too was rescued from a spoil heap, and had come perilously close to being lost. After some inspection Woodward almost threw it away as a mere bit of ironstone. He does not say what stopped him.

In August, during one laborious three-day weekend at the pit, three pieces of the skull's right parietal wall were recovered. All three were salvaged by the alert Dawson, one piece each day. Their discovery was the result of patient hours of poking through a mound of discarded soil or, as Dawson described it, "a heap of soft material rejected by the workmen." Before the excavations were closed down for the season in September, no more bones were found, though several flint implements turned up, paleoliths that had obviously been used as primitive tools. One flint was found by each of the three men.

Also uncovered were a few fossil teeth of various ancient animals or earlier forms of existing ones: mastodon, beaver, hippopotamus, horse, and primitive elephant, along with a short length from the antler of an extinct red deer. All these would be vital in supplying an age for the gravel, helping to date the hominid bones.

Long before Woodward had completed reconstruction of the skull, or the necessary preparations had been made for Piltdown Man's debut in the scientific world, rumors were flying. While the three men had agreed not to divulge to fellow scientists anything specific beforehand, and to give nothing at all to the newspapers, both Woodward and Dawson did talk in general about their finds with close friends and colleagues. As early as the start of October a letter reached Woodward from Manchester written by his old college mentor, Boyd Dawkins, praising and commenting on "your wonderful find." Dawkins' informed remarks regarding what he called "the absence of hollows for the reception of the convolutions of the brain" on the skull's inner surface indicated that Woodward had been writing him in considerable detail about all that was happening.

Another colleague, E. R. "Ray" Lankester, a leading zoologist who had held the chair of anatomy at Oxford and afterward served as keeper at the Natural History Museum, made no effort to disguise his joy. "I have been thinking of nothing else but that splendid human fossil all day. Ought you not at once to see about getting, say 500 pounds from the government grant reserve so as to start future digging and secure the rights . . . It seems to me it would be worth spending *thousands* on this, a regular systematic and complete sifting of every hatful of gravel in the neighborhood . . . I feel sure the whole family are there leg bones and all of a dozen individuals!"

Sir Arthur Conan Doyle, at his home, Windlesham, in Crowborough, was another who heard about and was immensely stirred by the rumors. By now in Sussex archeological circles, talk of the great new find was circulating freely, vastly quickened after the first news story broke in *The Manchester Guardian* on November 21. Doyle's new novel, *The Lost World*, with its strikingly original and detailed portrait of apemen and dinosaurs surviving in South America, had just been published (after running serially in *The Strand* magazine), and he was eager to know more of Piltdown's real apeman. Late in November he wrote Dawson offering to put his automobile at his friend's disposal (cars were not plentiful then in Sussex, and Doyle must have been aware that Dawson did not own one). Whether Dawson took advantage of the offer is not known, but he duly reported it to Woodward: "Conan Doyle has written and seems excited about the skull. He has kindly offered to drive me in his motor next week anywhere." Some two weeks later, if not before, Doyle paid a personal visit to the Barkham Manor site, though unfortunately he left no comment regarding his inspection of the pit.

The news also quickly leaped the Atlantic, bringing from the Smithsonian Institution an urgent inquiry as to the truth of the wild rumors then racing around Washington about a "remarkable" fossil skull newly discovered. In England, many of those who heard the news, especially scientists in and around London, were anxious to see and handle the specimens, but almost all were met with polite refusals. Apparently no more than two or three men outside the staff at the museum were given the privilege of inspecting the bones prior to their official presentation. One of these was Arthur Keith, renowned anthropologist, anatomist, and director of the Hunterian Museum at the Royal College of Surgeons.

Returning to London from a family vacation in Scotland, Keith somehow immediately heard the rumors. Impatiently he waited, expecting Woodward to contact him, as a courtesy if not for help and information. After several weeks of silence he could contain himself no longer, and on November 2 he wrote confidently to Woodward asking for "a glimpse of that wonderful find," adding in his delight, "It bucks me up to think that England is coming up trumps." Could Woodward drop a postcard setting a date for his visit? For Keith it required only a quick trip across town.

Somewhat later Keith was to admit that he and Woodward had never been more than professional acquaintances, and were not close even in that sphere ("In our chance meetings he had struck me as a proud and cold man, one with whom I found it difficult to establish a friendship"). He would also later confess that all the tantalizing rumors in the city concerning the great discovery had left him feeling very unhappy, in fact, nothing short of "jealous." It had been Keith's longtime professional hope to make the famous Hunterian Museum at the Royal College, as he said, the "proper home for all fossil remains of ancient man" in England, and he was quite conscious of having made a splendid start on his dream. Woodward's operation in South Kensington, on the other hand, Keith stated baldly, was not adequate for dealing with the challenge of Piltdown. The Natural History Museum "had an excellent anthropological collection, but no anthropologist, whereas I had trained myself for just such a task as had apparently fallen to a rival institution." The deficiency in South Kensington, he added frankly, lay precisely in the geology keeper's lack of credentials: "As a paleontologist Smith Woodward enjoyed, and deserved, the highest reputation, but he had no special knowledge of the human body."

Woodward's expertise with ancient fish, mammals, and reptiles equaled or exceeded that of any scientist in the field of animal paleontology, but he didn't know enough about human anatomy to be comfortable or even wholly competent as a paleoanthropologist. The charge, never made in public by Keith, was at least partially and technically true. Still, from the first, Woodward acted in complete confidence, displaying not the least hesitation in passing judgment on Piltdown's difficult mixture of simian-hominid morphology.

Because he was too busy for visitors—or perhaps he simply resented Keith's assumption of privilege—Woodward delayed for nearly a month setting a date for Keith to appear at the museum. At last he suggested December 2, barely two weeks before the meeting of the Geological Society, at which Piltdown Man would be unveiled. On the appointed day Keith arrived at Woodward's office and, though the time allowed him was surprisingly short, a mere twenty minutes, he found himself thoroughly fascinated by what he saw. Included was a model of the fully reconstructed skull with missing parts supplied:

> It was late in the evening when I arrived; the lights were being turned out in the great hall of the museum. When I met Smith Woodward in his private office, our mutual greetings were curt. He unlocked a drawer and laid the precious fossils on a table in front of me. I went over part after part, noting their massiveness, their complete state of fossilization, making a mental estimate as I went over them, of the probable size and shape of head and brain. Especially did I note the teeth and the ape-like formation of the region of the chin.
>
> The simian character of the lower jaw surprised neither of us; if Darwin's theory was well founded, then a blend of man and ape was to be expected in the earliest forms of man. I was shown the reconstruction that had been made, giving Piltdown man the large canine teeth of an anthropoid ape. The hinder part of the skull, to my passing glance, seemed to be wrongly put together. I left the museum in no doubt that a discovery of the highest importance had been made in the unlikeliest of places—the Weald of Sussex.
>
> On reaching Highbury [his home] I went to my diary and noted the various points that had surged through my head during those twenty minutes I spent at South Kensington.

The diary entry Keith made that night is still available. It is quite lengthy, detailing various measurements of the main fragments, along with specific commentary on individual pieces of bone. Among those aspects of the reconstructed skull that had "surged" through Keith's head were a number of specific points relating to both brain case and jaw which he felt sure were "wrong," errors he intended to pursue and expose. He had made good use of his twenty minutes.

With the start of December all of scientific London eagerly anticipated the evening of the eighteenth, when the Geological Society's regular meeting would be held, as usual at Burlington House in Piccadilly. The conference room always used for these meetings was arranged in parliamentary form, the chairman sitting at one end, while leather-covered benches for members ranged along both sides facing one another. It was not a large room, and more requests for seating than usual had been received, so that when the night arrived the room was filled beyond capacity, spectators squeezing in wherever a little space appeared. The meeting had "never been exceeded in interest or brilliance," reported the *Saturday Review*, ". . . never has the meeting room been so crowded. For the problem of our ancestry, and in particular our relation to the apes, appeals to every class of intellect." Few topics, the *Review* correctly added, "have in days gone by aroused more bitter controversy." Conspicuous in the crowd were the Archbishop of Canterbury, the Duke of Bedford, and Lord Avebury, one of the pioneers of English paleontology.

On a table at the room's center lay a model of the reconstructed skull, its appearance strangely mottled because the white plaster of the conjectured parts contrasted strongly with the dark brown of the original bone. Beside the skull were several boxes in which the actual bones themselves rested on soft cotton padding. At eight o'clock the evening's regular business was disposed of, and Charles Dawson was introduced. In the hush he rose and made his way to a place at one end of the long table, just in front of the chairman's dais. Standing near him was a slide projector aimed at a nearby screen. Before him stood a rostrum holding the manuscript of his paper, which had been prepared in collaboration with Woodward. Facing the distinguished audience, he appeared relaxed and confident.

It had all started "several years ago," Dawson began rather vaguely, and he went on to describe the events that had preceded Woodward's entry into the picture. As he talked he was careful to preserve the secret of the actual site at Barkham Manor, since no one wished to have hordes of the curious descending on the unfinished excavations. This reticence his audience understood.

He had been walking alone one day, he recalled, on a farm road close to Piltdown Common in Sussex. Occasionally he glanced down as he walked, and at one point he noticed embedded in the road under his feet "some peculiar brown flints not usual in the district." Surprised to learn, on inquiry, that the flints had not been hauled in from a distance but had come from a gravel bed on that very farm, he went in search of the bed. When he found it, two men were busily shoveling out the chunks of flint used for repairing the neighborhood's roads: "I asked the workmen if they had found bones or other fossils there. As they did not appear to have noticed anything of the sort, I urged them to preserve anything they might find. Upon one of my subsequent visits to the pit, one of the men handed to me a small portion of an unusually thick human parietal bone. I immediately made a search, but could find nothing more, nor had the men noticed anything else."

His many later searches at the site, said Dawson, still omitting to mention dates or time sequences, had yielded nothing. But then at last in the previous autumn, that of 1911, "I picked up, among the rain-washed spoil-heaps of the gravel pit, another and larger piece belonging to the frontal region of the same skull, including a portion of the left superciliary ridge." (This was the fragment containing the small bit of telltale human brow.) Having once examined a cast of the famous Heidelberg jaw, he explained, and remembering its unusual thickness, he at once noted in some considerable excitement that "the proportions of this skull were similar." Early that year, 1912, he had taken the bones to Dr. Woodward, and since then both had been hard at work on the excavations.

For so important an event it was a somewhat bare recital, though no one at the meeting registered a complaint. All present no doubt took it for granted that fuller details on these earliest aspects of the discovery would eventually reach print, and this is what happened. In the August 1913 issue of the *Hastings and East Sussex Naturalist* Dawson had a lengthy article entitled "The Piltdown Skull," covering some ten pages with illustrations. The picture he drew there about the events preceding his approach to Woodward is much more graphic and circumstantial:

Many years ago, I think just at the end of the last century, business led me to Piltdown, which is situated on the Hastings beds and some four or five miles north of the line where the last of the flint-bearing gravels were recorded to occur. It was a Court Baron of the Manor of Barkham at which I was presiding, and when business was over and the customary dinner to the tenants of the Manor was awaited, I went for a stroll on the road outside the Manor house. My attention was soon attracted by some iron-stained flints not usual in the district and reminding me of some Tertiary gravel I had seen in Kent.

Being curious as to the use of the gravel in so remote a spot, I enquired at dinner of the chief tenant of the Manor where he obtained it.

Having in remembrance the usually accepted views of geologists above-mentioned, I was very much surprised when I was informed that the flint gravel was dug on the farm and that some men were then actually digging it to put on the farm roads, that this had been going on so far as living memory extended, and that a former Lord of the Manor had the gravel dug and carried some miles north into the country for his coach-drive at "Searles." I was glad to get the dinner over and visit the gravel pit, where, sure enough, two farm hands were at work digging in a shallow pit three or four feet deep, close to the house.

The gravel is an old river-bed gravel chiefly composed of hard rolled Wealden iron-sandstone with occasional sub-angular flints. The men informed me that they had never noticed any fossils or bones in the gravel. As I surmised that any fossils found in the gravel would probably be interesting and might lead to fixing the date of the deposit, I specially charged the men to keep a look out.

Subsequently I made occasional visits, but found that the pit was only intermittently worked for a few weeks in the year, according to the requirements of the farm roads. On one of my visits, one of the labourers handed to me a small piece of bone which I recognized as being a portion of a human cranium (part of a left parietal) but beyond the fact that it was of immense thickness, there was little else of which to take notice.

I at once made a long search, but could find nothing more, and I soon afterwards made a whole day's search in company with Mr. A. Woodhead, M.Sc., but the bed appeared to be unfossiliferous. There were many pieces of dark brown ironstone closely resembling the piece of skull, and the season being wet, any fossil would have been difficult to see.

I still paid occasional visits to the pit, but it was not until several years later that, when having a look over the rain-washed spoil-heaps, I lighted on a larger piece of the same skull which included a portion of the left supra-orbital border. Shortly afterwards I found a piece of a hippopotamus tooth.

In the meantime there had been a revival of the study of early man, and I had lately had the opportunity of examining a good cast of the famous Heidelberg jaw. And so it came to pass that one morning I walked into the Natural History Museum to call on my old friend, the Keeper of the Geological Department, Dr. A. Smith Woodward . . .

Neither of Dawson's two published accounts of how his discoveries began mention a date. But he must have given one on the night of the eighteenth, either during the meeting or to reporters afterward, for the newspapers the next day in their lengthy accounts are specific. In place of Dawson's published references to "several years ago," and "many years ago . . . just at the end of the last century," the newspaper articles clearly state "four years ago," meaning 1908, a precision that could have come ultimately only from Dawson.

Continuing his presentation, Dawson next sketched at some length the geological character of the Piltdown area, enumerating its various soils, then listed the fossil finds made during that summer. The fragmented condition of the bones, he explained, had resulted from the regrettable breaking of what had originally been a more or less intact cranium: "Apparently the whole or greater portion of the human skull had been shattered by the workmen, who had thrown away the pieces unnoticed. Of these we recovered from the spoil-heaps as many fragments as possible." His recalling how the workmen mistook the skull for a "cocoanut" appears in the next day's news accounts, though not in the published text of his talk. Perhaps it was mentioned as an aside, or told to reporters afterward.

The jaw, Dawson went on, had been found in "a somewhat deeper depression of the undisturbed gravel." So far as he could judge, this was the very same spot in which "the men were at work when the first portion of the cranium was found several years ago." The important piece of occipital bone found by Dr. Woodward, he added, had turned up "within a yard" of the same spot, and at the same level. Only the jaw's right half was present, the break having occurred "at the symphysis" (chin area). The edges of the break had been abraded by contact with other hard objects as the jawbone moved along the bed of the river that had once flowed through the area. The cranial fragments, on the other hand, "show little or no sign of rolling or other abrasion."

To determine the extent of the skull's fossilization, said Dawson, a chemical test had been made on one of the cranial pieces. Performed by the public analyst for East Sussex, Samuel Woodhead of Lewes, the tests definitely showed that "no gelatine or other organic matter is present. There is a large proportion of phosphates (originally present in the bone) and a considerable proportion of iron. Silica is absent." This indicated virtually complete fossilization, all trace of the original animal matter having been transposed.

His part of the evening Dawson closed with an effort to assign an age to the skull according to its geological matrix. While the Piltdown gravel beds were undoubtedly of the Pleistocene epoch, he explained, in their lowest stratum were to be found "animal remains derived from some destroyed Pleistocene deposit probably situated not far away, and consisting of worn and broken fragments. These were mixed with fragments of early Pleistocene mammalia in a better state of preservation, and both forms were associated with the human skull and mandible, which show no more wear and tear than they might have received *in situ*."

From these and other facts, he concluded, it seemed that both brain case and jaw could not be placed earlier than "the first half of the Pleistocene epoch. The individual probably lived during a warm cycle of that age." Piltdown Man had roamed England, it would appear, if not a million years before, then certainly no less than half a million.

Woodward's presentation on this occasion was much the lengthier of the two, and far more technical. He opened with a minute description of each of the nine pieces of cranial bone, elaborately explaining how they related to one another according to what was known of skull structure in humans. The Piltdown cranium was altogether normal, he said, showing no trace of disease or of exaggerated growth pattern. Its reconstruction, as a result, while time-consuming, had been rather straightforward, the only difficulty being the absence of sufficient contact between left and right sides. Admittedly this left a ragged, tantalizing gap through the skull's middle from front to back, but sufficient indicators were present—traces of the midline, marks of various sutures—to permit a quite dependable reassembling.

Beyond question, said Woodward, was the fact of the cranium's essential humanity. The high forehead and rounded vault had held a brain measuring at least 1,070 cubic centimeters, about equal to the smallest brain found among modern primitive races. Most remarkable was its extraordinary thickness, at places reaching twelve millimeters, where the ordinary modern skull seldom exceeds five or six. Even more curious, this excessive thickness was not spread uniformly through the skull, but tapered from front to back. Pronounced at the rear, it diminished along the sides and was thinnest at the frontal region, though still exceptional. Further, the thickening was all due to expansion of the *diploe*, the honeycomb of bone tissue between the relatively thin outer and inner plates (or upper and lower), the covering layers of solid bone (Woodward called them "tables").

But about the cranium's basic human character there was no doubt, a fact confirmed by any number of clear indices. There was the unsymmetrical cerebellum and the very deep glenoidal cavity. There was the protuberance of the external occipital below the *tentorium*, and the downward inclination of the long axis of the ovoid opening at the external auditory meatus. Most significant, there was the absence of a *spina glenoidalis*, so common in apes.

The remarkable jawbone, however, exactly reversed this situation, showing itself to be almost wholly simian. Any number of features pointed to it, particularly the absence of the necessary ridge for anchoring the mylohyoid muscle, a ridge conspicuous in humans. In fact the jaw closely or exactly resembled that of an ordinary chimpanzee. It was only the dentition, undeniably human, that definitely signaled its true nature, linking it to the cranium.

Unfortunately the articular condyle—the little knob at the top extremity of the jaw's *ramus*, the ascending portion—had been broken or worn off. Here was the most regrettable loss of all, for this feature was the sole working contact between cranium and jaw, the swivel point for opening and closing the mouth. In man and ape it took quite different shapes. Had the condyle been present in the Piltdown bone it would at once have settled the question of whether cranium and jaw could have been rightly joined. In the absence of the condyle, the two molar teeth became of paramount importance. The pattern of wear, so obvious in their flat surfaces, connected them firmly to man. So marked and uniform a flatness "has never been observed among apes," Woodward explained. But the pivotal question regarding the teeth concerned the canine, missing along with the chin region in the Piltdown jaw.

Was the canine to be restored as it appeared in apes, large and prominent? Or should it be, as in man, smaller and less conspicuous? On this question Woodward compromised. It was quite reasonable to conclude, he said, that the missing portion of the Piltdown jaw would have had a fairly massive forward thrust. To help fill this extent of jaw would take a canine larger than ever appears in man. But its apex must be angled forward since it could not have risen much above the distinctive molars behind it, both of them ground to a perfect flatness. The chewing action of man and apes differs fundamentally. Unimpeded by stout protruding canines, the human jaw is free to move laterally, as well as up and down. But for the ape, side-to-side movement in the jaw is severely restricted by the dominating canine. The two-way grinding action that eventually leveled human teeth is absent from an ape mouth.

Should the creature represented by the Piltdown skull, asked Woodward in closing, be considered a new species of Homo sapiens, or was it a hitherto unknown genus? His answer needed some explanation.

A million years back, if not two, he said, the earliest forms of man spread through western Europe "were already differentiated into widely divergent groups." As to which of these roving bands gave rise to modern humans, Woodward was in little doubt. It was not Heidelberg, he declared, which must finally be judged "a degenerate offshoot of early man," one of those evolutionary lines that had long ago become extinct. The same verdict covered other possible hominid branches as well, such as the Neanderthals and England's own problematic man of Galley Hill. In that case, Woodward concluded, the human race might well "have arisen directly from the primitive source of which the Piltdown skull provides the first discovered evidence." While he did not use the term, he meant that Piltdown Man was the true missing link, that elusive halfway creature talked about and looked for since Darwin's day.

A proper name for this new member of the family *Hominidae*, Woodward suggested, should reflect its status as a beginning, thus it might be called *Eoanthropus*, the dawn man. Allowing due honor to its discoverer, the particular species could be designated *Eoanthropus dawsoni*.

The evening's third speaker, Grafton Elliot Smith of Manchester University, the most eminent neurologist in England, had been brought in by Woodward almost at the last minute. He was to provide a specialist's view on aspects of the Piltdown brain, as indicated by the skull's endocranial features (the markings on the inside surface), studied by means of a cast. Smith's presentation was brief, lasting hardly ten minutes, but it strongly reinforced all that Woodward had proposed.

The Piltdown brain, announced Smith, was more primitive than any human brain he had ever seen or heard of. At one and the same instant, it was "the most primitive and the most simian brain so far recorded." Of its several striking aspects, perhaps the most noteworthy concerned the "deep excavation of the temporal area, to form the wide bay between the inferior temporal pole and the cerebellum." Just here it was that he saw unmistakable, even exciting indications of the brain's later expansion, causing "the very different configuration that it represents in modern man." This was a fortunate alteration, indeed, he said enthusiastically, for in the end it had directly generated man's wonderfully expanded powers of memory and speech.

No real paradox was involved, Smith assured his listeners, in this combining of what might seem contradictory elements in one individual. That a human brain should be found topping a beast's muzzle was no cause at all for scholarly doubt or hesitation. In evolutionary progress "the growth of the brain preceded the refinement of the features and of the somatic characters in general." In England, as his audience well knew, Smith had long been the foremost advocate of the brain-first hypothesis.

The audience discussion that followed the presentation was a lively one, in which not all the aroused scientists were ready to accept *Eoanthropus dawsoni* quite on Woodward's terms. Nine speakers rose to offer opinions. All were distinguished names in the fields of geology, anatomy, paleontology, and anthropology, and several had previously inspected the skull, as well as the gravel beds at the excavation. Two topics were prominent: the age of the specimens and the association of the jaw and the cranium.

On the question of age, out of nine speakers five called for the decision to be postponed while awaiting further data. Of the remaining four, one declined to express an opinion on age, and another fully agreed with the stated figure. Two felt strongly that it should be pushed even farther back, as far as the late Pliocene.

As to the claim that jaw and brain case were one, forming a single individual, the response was similar. Only four out of the nine were at all comfortable with the idea, while one was uncertain, and three held off entirely from giving an opinion. One man, however, anatomist David Waterston, spoke forcefully and in detail against the notion that the two parts added up to one whole skull. The brain case, he said, "was human in practically all its essential characters," while the jaw with equal certainty "resembled in all its details the mandible of the chimpanzee." Here was an emphatic divergence of identity, making it difficult to believe that the two specimens had ever been united in the same individual. With that, Waterston suggested taking a more searching look at one small item in Woodward's portrait of the cranium. Woodward, he said, had just now pointed out how closely the temporal bones of Piltdown, with the glenoid fossae, resembled the corresponding parts in modern man: "It must be borne in mind that the configuration of the glenoid fossae in man was such as to adapt them for articulation with a human jaw, and not with the mandible as found in the chimpanzee. If the jaw had formed part of the skull, it was precisely in the temporal bone (glenoid fossae) that one would have anticipated some variation in structure from the present-day condition."

The socket (glenoid fossa) in the Piltdown cranium, Waterston was saying, had obviously been intended to receive a *human* condyle, as Woodward admitted. But if the two specimens were to be joined, this produced the drastically untypical situation in which the chimpanzee jaw, while remaining simian in all other respects, must have evolved a completely human condyle, fitting into a socket recess in the cranium, which remained wholly unadapted. (Shortly afterward, Waterston put his feelings against linkage in simpler terms. It was like joining "a chimpanzee foot with an essentially human thigh and leg.") The missing condyle, so unfortunately broken off, would continue for some time to underlie the main disagreement.

Arthur Keith also offered some remarks—curiously brief and circumspect in view of what was to come—in which he managed to mingle both support and disagreement. Great skill, he conceded, had been used in reconstructing the skull, yet he felt sure that a significant portion of it was not quite right. The restored chin region with its supposed incisor and premolar teeth, and especially the projecting canine, had been made far too apelike. In his view, everything pointed to the skull's having much more the appearance of modern man (in line with his own favorite theory of the vast antiquity of present humanity). Still, Keith emphasized, Piltdown, whether more or less bestial, was beyond doubt the most significant fossil discovery yet made in England, equal in importance to any fossil ever found in any country.

As the last of the nine finished speaking, Woodward and Dawson were invited by the chairman to respond. First Dawson came forward, but he chose not to dispute any of the speakers and offered only some remarks on the problem of dating ("an earlier date for the origin of the human remains" would not disturb him). Woodward's equally short reply concerned his own conjectural restoration of the chin region and the prominent canine. These, he admitted, could be seen as "a bold experiment." Further digging at Barkham Manor, he announced, was planned for the next summer, and perhaps other bones would turn up to eliminate any lingering doubts.

In a session with reporters afterward, Woodward allowed himself to unbend a little, becoming more positive. Long before Piltdown came to light, he said, and on quite other evidence—mainly the record of flint tools—a primitive race of men had been identified as having lived in England many thousands of years in the remote past, even before the "known cave dwellers." Until then, no relics of this earlier race had been known. But now the Piltdown remains wonderfully established the truth of the old idea. "Our discovery," he stated almost exuberantly, "confirms in a striking manner the theories of science."

The next day, and for a week thereafter, newspapers in England and most other countries blared to readers the sudden arrival on the world's stage of ape-jawed but clear-eyed Piltdown Man, the missing link. FIRST EVIDENCE OF A NEW TYPE, a headline in the London *Times* proclaimed.

3

Challenging the Skull

The more he stared at, handled, and measured the cast of the Piltdown skull sitting on the table before him, the more was Arthur Keith convinced that the entire fascinating assemblage had been incorrectly put together. In his eyes, the model represented nothing less than "a grave blunder," with the blame laid squarely on the shoulders of the inexperienced Woodward. In addition to the jaw being much too apelike, the brain case had been shrunk, egregiously depriving the brain of its true capacity. The height of the left parietal wall, even by itself, indicated as much. To restore the skull's true original conformation would require a fairly drastic overhaul.

At stake, Keith decided, was much more than a merely technical tug-of-war between rivals in paleontology. Proper assembly of the skull would show that modern man—crucial aspects, at least, of his brain and body—had begun to take shape much farther back in the abysm of time than anyone had imagined. Nothing less than the whole fundamental approach to the fledgling concept of human evolution was in the balance.

The cast of the restored skull, with separate casts of each of the bone fragments, had been made by a commercial firm under contract to the museum for distribution to interested scientists. It had reached Keith's office in mid-May 1913, and by the first days of June he felt that he had uncovered Woodward's basic error: the skull's midline had been wrongly identified. Needed corrections would shift the entire left part of the cranium away from the center, increasing the cranium's height by more than half an inch, causing the entire dome to be stretched in width and fullness. In anatomical terms the enlargement was breathtaking, adding no less than four hundred cubic centimeters to the Woodward version.

Also changed was the marked lack of symmetry between the brain's two halves, now entirely eliminated. An asymmetrical condition was a distinctive feature of the modern brain, one hemisphere becoming dominant. All ancient skulls revealed a sameness or close similarity between left and right, a definite symmetry, with both hemispheres in balance.

Woodward's mistake in the brain size, Keith judged, was at least partly a result of his drastically skewed ideas about the excessively primitive nature of the jaw, its simian complex. The lack of a strong chin, squared off, and especially the presence of large canines, in the fashion Woodward had imagined, could not be defended. For one thing, they pointed to a curiously contradictory creature whose brain equaled in size that of a modern human, but who was physically unable to speak. It was the capacious human chin, in Keith's view, that provided the interior space needed for the free and supple movement of the tongue in speaking.

News of the full scope of Keith's radical reconstruction of the skull first reached Woodward through the neurologist Grafton Elliot Smith, who was on friendly terms with both men. In London on business early in July 1913, Smith visited the Royal College in Lincoln's Inn Fields, where he found Keith in his office at the Hunterian Museum. Promptly the next day he sent Woodward a letter of warning:

> I saw Keith and he showed me his restoration of the Piltdown skull . . . As the matter of the putting together of the fragments vitally affects my part of the work I carefully examined all the points raised by him; and as a result I am quite convinced that we shall have to modify the restoration in some respects . . . the ridges of separation of the cerebral and cerebellar depression on the occipital bone are *not* in the mesial plane . . . the occipital in your reconstruction must be rotated to the right. This will throw out the right parietal and broaden the skull. . . .

According to K. the temporal bone has been inclined at an angle with the parietal. His third point is that the mandibular fragment [the jaw] includes the symphysial region; so that normal sized teeth will completely fill up the space allowed by the alveolar process. I have convinced myself that the occipital and right parietal regions in your reconstruction need modification. . . .

On receiving this letter, the anxious Woodward promptly wrote Keith a request for a meeting to discuss the issue. It should be a private meeting, he suggested, perhaps including a few colleagues on either side, but otherwise kept well back from public view, and especially from the newspapers, for the good of science. To this Keith agreed, but Woodward's hopes for keeping the dispute a private matter were soon defeated.

The first meeting of the two, held at Keith's office on July 10, brought no change in Woodward's opinion, but it was a result due in large degree to personality differences. In laying out his ideas, the confident Keith had been "irritable" in manner, almost abrasive. In turn, the sensitive, defensive Woodward had not hidden his own growing annoyance at the questioning of his work. Neither paid heed to the practical advice of a colleague present at the meeting to confine their attention to scientific measurements ("If you fellows would stick to the solid thing on the table and not wrangle about fancies, all would be well"). Woodward left Keith's office that afternoon more than a little "upset."

It was not a month later that the rancorous disagreement between the two well-known scientists found itself elevated to the status of major news. The International Congress of Medicine, scheduled for London in August, got wind of the dispute and invited the men to air their differences before its large body of distinguished experts so that the representative group might offer its own judgment in the matter. Both accepted, Woodward because there was no legitimate way to refuse, Keith because he wanted a platform.

On the day set for the dual appearance, the London *Times*, in announcing the unusual confrontation, also made an effort to clarify the issue for its readers. "If Dr. Smith Woodward is right," the account explained, "we have to seek the beginnings of our modern culture and civilization at the middle of the Pleistocene period; if his opponent's reconstruction is well founded, we have to go a whole geological period further back—perhaps a million of years—to find the dawn of modern man and his culture."

On the afternoon of August 11, an impressive group of scientists, perhaps as many as forty, some just arrived from the Continent, converged on the Natural History Museum in Cromwell Road. Welcomed by Woodward in the museum's lecture hall, for two full hours they listened patiently to his calm, detailed demonstration, interrupting now and then with brief questions. At the session's close the chairman, Professor Thomson of Oxford, said that all discussion should be delayed until after Keith's counterdemonstration.

Across town went the group, transported in a fleet of carriages, buses, and motor cars to the Royal College. There they were conducted to the lecture room, where, as Keith took the stage, they settled down and grew silent.

Perched on a table in view of all were several skull reproductions, with a scattering of separate casts of the jaw and the other small fragments of bone. Conspicuously propped against the Piltdown skull model, as envisioned by Keith, was an identifying label. It read not *Eoanthropus dawsoni*, the name applied by Woodward, but a new one devised by Keith, *Homo piltdownensis*. As all promptly realized, the change indicated a demotion for the creature, making it a species of Homo sapiens rather than a new genus. A surprisingly brash move, it was a clear sign that Keith meant business.

His tone distant and a little superior, Keith opened with the statement that in both animals and man there existed certain definite laws controlling the formation of skulls. One of these related to the groove found on the inside of the skull's roof, a groove caused by the great venous channel occupying the exact middle of the cranial chamber. No skull had ever been seen lacking such a groove. But in Woodward's model, he said, "This groove has been placed nearly an inch away from the middle line. When the bones of the side and roof of the skull were separated so as to place the venous groove in its true position" the result was a sudden and dramatic increase in brain size. The small brain calculated by Woodward, at just over a thousand cubic centimeters, was enormously different from one of nearly fifteen hundred cubic centimeters, Piltdown's true capacity as revealed "when the bones were *rightly* articulated."

Lifting the model skull from the table and holding it high, he added that another serious error in Woodward's version concerned the skull's base, where it met the spinal column. As imagined by Woodward, the Piltdown backbone "came so near the palate that the person, as reconstructed at South Kensington, could neither breathe nor eat." After a pause he went on, pointing to the skull in illustration, "It is possible that *Eoanthropus* could not speak. But we must suppose that he could breathe and eat! If a student had presented a skull like this one he would have been rejected for a couple of years."

Scarcely controlling himself, Woodward rose to his feet. The elements he had provided for the skull's base, he declared, were admittedly highly conjectural, certainly not intended to represent "the actual parts," but meant only to complete the remodeling. The icy words implied that Keith knew very well that the skull's base was not a legitimate point of contention. He was cheaply taking an unfair advantage.

"Then why represent them?" Keith cut him off brusquely. In reality, he went on, Woodward's initial error arose from his misunderstanding the true nature of the Piltdown jaw and teeth. A chimpanzee-type jaw and palate had been mistakenly fitted to a brain case that could not carry them. "All the evidence was against a big canine tooth," Keith stated firmly. "There was no room for a great anthropoid eye-tooth; the chief muscle of mastication was smaller than in modern natives of Australia; the joint for the lower jaw was exactly as in modern man, and the mechanism of that joint was incompatible with a projecting canine."

Concluding, Keith insisted that the technique of skull reconstruction, then so new, especially of an incomplete specimen, could be accepted as thoroughly reliable. In fact it was a discipline in itself, a veritable science with rules of its own. Given certain parts and fragments, the correct modeling of an individual whole skull must follow as surely as if it were "a problem in Euclid." The implication was plain: only a professionally qualified human anatomist, one of long experience such as himself, could be expected to solve the puzzle of a broken and partial skull. Almost, it seemed, Keith was on the verge of declaring to the rapt audience that Dr. Woodward woefully lacked this qualification.

In the general discussion that followed, the various speakers were careful to spare the obviously chagrined Woodward any further embarrassment. But it was clear to all that the distinguished jury was ready to award his opponent the victory. Keith's version of the skull appeared the truer one, they concurred, based as it was on reason and plainly stated scientific principle. As the London *Times* reported, at the meeting's end the chastened Woodward was moved to make a concession of sorts, but only as regarded the shape of the cranium. "It was possible," he admitted, "that in the original construction some error had been made in the articulation of the bones, a malposition which might diminish the actual size of the brain."

But regarding the primitive aspects of the jaw and its large, prominent canine, Woodward held his ground, if not with quite his old assurance. The official designation for Piltdown Man, he insisted, *Eoanthropus dawsoni*, was fully justified by the apelike jaw alone. On that legitimate basis, if no other, it should be retained, and he would accept nothing less.

His stand was a timely one, not to say fortunate: less than a month later Keith's triumph suddenly evaporated and Woodward found himself fully vindicated. Against all reasonable expectation, the canine in question had been found. It was large and apelike. It had been discovered in the same section of the Piltdown diggings that had yielded the jawbone. Advance information said that in its size, shape, and wear pattern it matched exactly the canine predicted by Woodward's model.

As soon as the weather cleared in the spring of 1913 the excavations at Piltdown were resumed. Starting in late May, most weekends saw both Dawson and Woodward on hand, with the laborer, Hargreaves, doing the heavy work. Through the whole season there occurred only one interruption of note, but it was a large one, though occupying only a single day and planned well in advance. It was an occasion that marked the opening of the Piltdown site to all qualified observers—in effect to public viewing. An unusual if not exactly surprising development, it produced one strangely unexpected result, which could not have been welcomed but might sensibly have been anticipated.

On July 12, a Saturday, a large group of geologists from London, with wives and children, was put aboard three buses and taken on a tour of the Lewes-Piltdown countryside under the guidance of Dawson and Woodward. After viewing the area's many hills and valleys—debarking at certain points while Dawson sketched its geological history—the party finally arrived at Barkham Manor. There it was greeted by the manor's tenant, Robert Kenward, and by a second large group of geologists, amateur and professional, assembled from Sussex. Standing at the pit's edge, Woodward faced a sprawling audience of at least a hundred men and women. Briefly he described the pit's soils and gravels, then followed with a short history of the fossil discoveries. In closing his talk, as one paper reported, he actually invited his listeners to make a personal inspection of the site, or seemed to: "He pointed out where the skull was found, and the party examined the spot closely, and the ground in the immediate vicinity."

That this mob of visitors was given access to the pit itself seems, incredibly, to be a fact. A photograph of the occasion shows the crowd massed at one end of the pit, with at least two persons standing at the bottom, and two more about to descend. *The Sussex Daily News* added the astonishing information that some of the visitors were actually allowed to dig in the pit unhindered: "Taking advantage of the extended halt many of the party proceeded to carry on excavations, but no more skulls or indications of their presence were discovered."

Neither Woodward nor Dawson, nor anyone else, it appears, made the least move to halt this random searching, nor did either of them ever make a reference to it afterward. At the ceremony's end, the entire party repaired to the Uckfield town hall where an elaborate tea had been prepared. There the gathering inspected "a cast of the skull, while photographs of it found a ready sale."

Also frequently working at the pit this second season, having returned from France to continue his studies at the Jesuit house in Hastings, was the young priest, Teilhard de Chardin. He arrived in England at the start of August, and after a week of parish service in Canterbury he stopped at Lewes for a few days, where he was the guest of Charles Dawson at his home, Castle Lodge. On the weekend of August 9 he was at the pit, helping with the endless labor of sifting, and was present when Dawson happily picked up several thin slivers of bone, later identified as pieces of the Piltdown nose. This time it was not the spread-out gravel of the spoil heaps that yielded the tiny specimens, but gravel that had already been somewhat loosened in the pit bottom. They were spotted by Dawson, as he recorded, while he stood and watched Hargreaves shoveling:

> While our labourer was digging the disturbed gravel within 2 or 3 feet from the spot where the mandible was found, I saw two human nasal bones lying together with the remains of a turbinated bone beneath them *in situ*. The turbinal, however, was in such bad condition that it fell apart on being touched and had to be recovered in fragments by the sieve; but it has been pieced together satisfactorily by Mrs. Smith Woodward.

An infrequent visitor to the excavation site, Maude Woodward's presence that day was to prove helpful. Their own hands and fingers, the men agreed, were too thick and clumsy for the task of lifting the delicate pieces of nose bone and the shattered turbinal from the rough gravel. As Dawson remembered, Mrs. Woodward's hands were small and graceful, her fingers especially slender, so they asked her to pick up the various bits of bone and put them in a box. This she did, and later on her own managed to glue the tiny pieces of turbinal back together.

When Dawson says that he "saw two human nasal bones" *in situ*, of course, he does not mean that he recognized the bones' true character the moment he sighted them embedded in the soil. That impression of his words comes from the telescoping practices of scientific writing (and one of its inadequacies). He means that *something* in the soil being worked by Hargreaves caught his eye, which on close inspection appeared to be human nose bones, a judgment that was later confirmed by analysis. The delicate turbinal bone (a spongy swirl from the nasal passage) especially would have required some effort to repair and study. The small, "gently arched" nasal bones, it was decided, came from a face that was unusually large, resembling those of "low races of men."

After the August 9 weekend, Teilhard's duties at the seminary, including a ten-day spiritual retreat, kept him away from Barkham Manor for the following two weeks. Then he was back at the pit again on Saturday, August 30, eager to be part of what he thought was to be the season's final dig. (In reality the work would continue through September.)

Early on the morning of the thirtieth, Teilhard left the seminary and took the short train ride to Lewes, where he breakfasted at Castle Lodge with Dawson, his wife, and stepson. Well before noon, he and his host were on a train speeding north to Uckfield, a fifteen-minute ride. There they met Woodward, who had that morning come down from London, and together they all proceeded by car to Barkham Manor. For some unexplained reason, the workman Hargreaves failed to show up, so that the three scientists took turns with the pick and shovel, loosening the hard-packed gravel and heaping it together for sifting through the boxed screens. As they worked, they had the amusing distraction of a large and assertive goose parading the grounds, which "never left us alone while we were digging," Teilhard later wrote. "At times he was gentle, at others cantankerous towards us, but he was always ferocious towards the passers-by."

An important part of the operations at the pit in 1913 called for particular attention to the spoil heaps, which had already yielded five fragments of the skull. These consisted of loose, extraneous soil from the different levels, lying at random around the borders of the elongated pit. Periodically brought together in a selected spot, the heaped gravel, now spread thin, would be left overnight, or for days or weeks, so that the frequent rains could wash it free of the disguising dirt. Compacted through the centuries, the dark soil clung tenaciously to everything, obscuring shape and texture and hiding the true colors and the shading of the light-to-dark-brown and blue-black gravel. Or, as Dawson explained about the operation on the thirtieth, "All the debris within five yards was sifted and washed, and afterwards strewn on a specially prepared surface." The surface was then "mapped out in squares," and each square minutely inspected and marked off.

It was while poking through one of these spoil heap squares that Teilhard found the crucial canine.

Of the moment itself, Teilhard had little to say, then or later. In a long, chatty letter to his parents (he was in the habit of writing them regularly), he mentioned the incident of the tooth but in a brief fashion devoid of specifics. "This time," he wrote, "we were lucky; in the earth dug up from the previous excavations and now washed by rain I found the canine tooth from the jaw of the famous Piltdown man—an important piece of evidence for Dr. Woodward's reconstruction plan: it was a very exciting experience! Imagine, it was the last excavation of the season! And so it was with a light heart that I returned to Hastings."

Dawson himself had even less detail to offer on the find than Teilhard. It was Woodward who took the trouble, within a year or so, to provide a somewhat fuller record, though he too omitted vital facts:

> For some time we had been making an intensive search for the missing teeth of the lower jaw round the spot where the lower half of this jaw was found. We had washed and sieved much of the gravel, and had spread it for examination after washing by rain. We were then excavating a rather deep and hot trench in which Father Teilhard, in black clothing, was especially energetic; and, as we thought he seemed a little exhausted, we suggested that he should leave us to do the hard labour for a time while he had comparative rest in searching the rain-washed spread gravel.

Very soon he exclaimed that he had picked up the missing canine tooth, but we were incredulous, and told him we had already seen several bits of ironstone, which looked like teeth, on the spot where he stood.

He insisted, however, that he was not deceived, so we both left our digging to go and verify his discovery. There could be no doubt about it, and we all spent the rest of that day until dusk crawling over the gravel in the vain quest for more.

A second, briefer mention of the incident by Woodward occurs in an interview with a London paper that he gave a dozen years later. It reinforces the fact that the discovery of the canine was the culmination of a conscious and deliberate effort, a focusing of attention on one item which, at that juncture, was eminently understandable. In the *Evening News* he is quoted as explaining: "We searched for the tooth in the same way as miners 'wash' for gold. But we didn't find it that way, after all. We found it at last—the tooth they said didn't exist—by spreading the gravel out on the ground and crawling over it examining every inch as we went. It was actually found by a Frenchman who was helping us in the search."

Some days after the canine find, Teilhard's assignment at the Hastings seminary was completed, and he left Sussex on his way back to France. Passing through London, he stopped for several nights in the city as a guest of the Woodwards ("He had me write my name on a piece of cloth covered with signatures of many geological celebrities," he wrote his parents, apparently unaware that he had been given a place on a table-cloth that was to become famous). He passed some days at the Natural History Museum studying fossils in general, and then he was gone.

Teilhard's role in the Piltdown story had ended. Before he had reason to return again to the Hastings-Lewes area, World War One had broken out, and he entered military service as a stretcher bearer. Afterward, the currents of his life in both science and religion, as well as philosophy and mysticism, took his attention far elsewhere (remarkably, he would be intimately connected with the discovery, in China in the twenties, of *Sinanthropus*, Peking Man, now among the pivotal fossils of evolutionary theory). In time, with such challenging and controversial books as *The Phenomenon of Man* and *The Divine Milieu*, he would earn fame for his ambitious attempt to fuse evolutionary theory with a religious and spiritual concept of man and his origin. But always he would think of his Piltdown days as "one of my brightest and earliest palaeontological memories."

Delighted with the turn of events, Woodward postponed announcing the new tooth only long enough to make a thorough study of it. Then at a regular meeting in September in Birmingham of the British Association for the Advancement of Science he gave out the news. (The effect was only a little dulled by the London *Daily Express* having released the first bare facts ten days earlier. For that account, Woodward had refused to do more than confirm the discovery, and to predict that it would prove "of tremendous importance.") With the Birmingham announcement there began a protracted public debate on Piltdown Man that was to last into the following year, growing ever more agitated.

The shape of the new tooth, Woodward told the British Association audience, "corresponds exactly with that of an ape, and its worn face shows that it worked upon the upper canine in the true ape fashion. It only differs from the canine of my published restoration in being slightly smaller, more pointed, and a little more upright in the mouth." Here was definite proof, he insisted, that the missing front teeth of *Eoanthropus* closely matched the simian pattern, upholding all his conclusions.

As for the purported "problem" of whether cranium and jaw belonged together, he felt that now there really was no possible doubt. The two molar teeth still in the jaw, worn flat by side-to-side chewing action, "are typically human." Also human were the jaw's muscle markings, and again, the jaw's being found in the earth very near the cranial fragments practically guaranteed their "natural association." The human-ape brain case definitely made one with the ape-human jawbone.

Feeling a new confidence, Woodward also rescinded his earlier, reluctant concession to Keith about the brain size. A new study he had made, he declared with satisfaction, showed that "the only alteration necessary in my original model . . . is a very slight displacement of the occipital and right parietal bones." It was quite a negligible adjustment, leaving the brain's capacity "nearly the same as that I originally stated." With that, he attacked Keith directly. It had only been by "distorting" the curve of the skull's underjaw, he charged, along with a number of other professional lapses, that his opponent had been able to reach such obviously false conclusions. Deliberate distortion seemed to be implied in his assertive tone.

The Birmingham presentation closed with Woodward proudly claiming his friend Grafton Elliot Smith as an ally once more. Though Smith had at first accepted Keith's ideas on the cranium, he had now inspected Woodward's new cast, and he "allows me to state that he finds it in all essential respects correct." Woodward's gratification with the positive effect of his Birmingham announcement is evident, an effect he made haste to clinch by prompt publication of his paper. Only three weeks later it was available in the monthly *Geological Magazine*. The arrogant Keith, he felt certain, faced with the canine and its many scientific implications, must now of necessity fall silent. But in this he had badly misjudged his adversary.

In *Bedrock Magazine* that October the resilient Keith presented his well-honed response. Projecting canine teeth, he repeated, were an impossibility in the Piltdown mouth. Rising as barriers, they prevented movement side to side, "such as is seen in animals which chew the cud, and also in man during normal mastication." Animals endowed with this sideways chewing motion *never* have prominent canines. Further, a sideways-moving jaw required a bone structure specially jointed to the cranium, with the knobs of the jaw thrusting upward into the cranium in a well-defined manner. In chewing, as the jaw is carried left, the right knob slips a bit from the socket into the "articular eminence." In the Piltdown skull that "eminence" was particularly well developed, "better even than in modern man." The conclusion was unavoidable that the Piltdown jaw did in fact move strongly side to side, in which case the canines definitely could not have risen higher than the other teeth. Yet here in this latest find from Barkham Manor was a tooth patently of the projecting type, a simian canine beyond a doubt.

In addition, commented Keith, the new tooth had other puzzling features "which raise a suspicion as to whether or not it does belong to this particular mandible." Its color was a much darker shade than the two molars still in place. Why was that? But far more striking was the disparity in age between jaw and tooth. Hard wear was traceable in the canine, the crown being especially worn down. But X-ray pictures of the jaw showed that the third molar, judging by the condition of the empty socket, had not yet fully erupted. Yet in apes the canine and the third molar erupted at the same time: "To me it seems an impossibility that the canine could be worn to such a degree and the third molar tooth not erupted in one and the same individual." One tooth reflected Piltdown as immature, the other as older, even aged.

But here, unexpectedly, Keith backed off a little. He had begun to make some experiments of his own, he said, to see whether there could have been some "peculiar arrangement" of the canine teeth that would permit side-to-side movements of the jaw "in the Piltdown race." If so, perhaps the big canine from Barkham Manor had belonged to someone else—say an older cousin of Woodward's man. The *Bedrock* article ended by resolutely restating Keith's original figure for Piltdown's brain capacity. It was not Woodward's ridiculously low figure, but a "massive" 1,500 cubic centimeters.

During the remainder of the year and into the next the controversy slowed but never stopped (a scholarly dentist, W. C. Lyne, at one point reported the strange contradiction he found between the canine's heavily worn condition and the apparent immaturity of its pulp cavity, only to be ignored). Especially in the pages of *Nature*, England's leading scientific journal, the fight went steadily on, the principals being Elliot Smith and Keith.

With detailed drawings—each being careful to call the other "my friend"—they battled point by point over highly technical aspects of skull morphology. Also raised was a new factor, one which would come to have importance, the question of whether a cast of the skull gave sufficiently minute and faithful detail to permit serious study (Smith thought not, Keith that it did).

Not content with print, Smith next carried the fight over into several talks he gave to scientific bodies, the highlight coming with an appearance in February 1914 before London's Royal Society. Mentioning Keith by name, he flatly condemned the great anatomist's published opinions on the Piltdown skull as woefully misinformed. Piltdown Man, he insisted in contradiction of Keith's well-known views, had certainly possessed an asymmetrical brain, certainly could talk, and certainly was the direct ancestor of modern man. "I had to speak straight," he told a friend afterward, "because so many British anatomists have been content to dance to Keith's rag-time, without attempting to think for themselves."

Not expressed at the Royal Society meeting but included in a letter written at about this time was Smith's private view of Keith's motive in so strongly opposing Woodward's ideas: "his game of deliberately fouling the pitch—and incidentally acquiring a little notoriety and hard cash." Keith's almost rabid ambition, Smith felt sure, was at the bottom of it all.

Keith himself was in the audience at Smith's Royal Society talk, and in the discussion that followed he went on the attack: "I did not mince my words," he later wrote. The ensuing exchange became downright unpleasant, and at its finish, as the worked-up crowd jostled its way out of the hall, it happened that the two antagonists were propelled near each other. "I shall never forget the angry look he gave me," recalled Keith later. "Such was the end of our long friendship."

The noisy, months-long debate between Keith and Smith settled nothing. As the year 1914 wore into summer the public discussion of Piltdown went on, in the privacy of laboratories and museums, now and then in print, the arguments revolving around the same, already much-disputed issues. Just as it began to seem that there was nothing more to be said, there came another surprise, as exciting in its way—"singular," Woodward called it—as the finding of the canine the previous year. Late in June at the Piltdown pit, a large bone tool, roughly fashioned from a part of the skeleton of some huge animal, came to light. This time Woodward himself was the main discoverer, and he happily described the find as unique, "unlike anything hitherto known among the handiwork of prehistoric or primitive men."

Until the new discovery, that summer's dig had yielded only small bits of two animal teeth, ancient rhinoceros and mastodon. It was while the regular shovelman, Hargreaves, was busily breaking soil in an untouched patch of ground—a spot from which the border of hedge had just been removed—that his pick hit something hard about a foot below the surface. At work nearby was Woodward, and at that precise moment he happened to be looking in Hargreaves' direction:

I was watching the workman, who was using a broad pick (or mattock), when I saw some small splinters of bone scattered by a blow. I stopped his work, and searching the spot with my hands, pulled out a heavy blade of bone of which he had damaged the end. It was much covered with very sticky yellow clay, and was so large as to excite our curiosity.

We therefore washed it at once, and were surprised to find that the damaged end had been shaped by man and looked rather like the end of a cricket bat; we also noticed that the other end had been broken across, and we thought it must have been cracked by the weight of the gravel under which it was originally buried.

Mr. Dawson accordingly grubbed with his fingers in the earth around the spot where the broken end had lain, and soon pulled out the rest of the bone, which was still more surprising. This piece was also covered with sticky yellow clay, but when we had washed it we found that it had been trimmed by sharp cuts to a wedge-shaped point.

The first piece taken out was a flat length of bone about ten inches by four inches. One end was a diagonal slash, an obvious break. The beveled condition of the other end was clearly the result of deliberate shaping: several rows of small, fine cuts across the edge had rounded it off nicely.

The second piece, the one dug out by Dawson, was shorter, some six inches in length, but it too was about four inches across, and it had a broken end that matched the break in the first piece. The shaped end was curious; it had been tapered by a series of delicate slices made by a sharp instrument. When whole and intact the implement, more than a foot long, would have served someone as a sturdy tool, good perhaps for grubbing up roots to eat, among other uses.

The yellow clay spread stickily on the bone's surface was a sign that its original resting place had been in the flint-bearing layer of clay at the pit's bottom, some five feet down. Only in one way, decided Woodward, could it have been brought to rest so near the surface, and that was by the careless shoveling of the busy road-menders. Most probably it had been "thrown there by the workmen with the useless debris when they were digging gravel from the adjacent hole . . . it agrees in appearance with some small fragments of bone which we found actually in place in the clay below the gravel."

The implement's two flat sides were of different textures, and this fact, together with the other structural markings, indicated its origin in the leg bone of a very large animal. Days later, some comparative study of fossils back at the museum soon uncovered its precise source: it was a thin slab chipped off the left femur (thigh) of a proboscidean, probably *Elephas meridionalis*, an extinct elephant larger than a mammoth.

Initial opinions on the implement differed wildly. From so early a period nothing like it had been seen before, and the questions mounted.

Had it been whittled when the bone was fresh? That would be crucial. If later, it lost considerable worth. The cuts had apparently been made by a skilled hand. Yet it appeared to some that they also could have resulted from the patient gnawing of an ancient beaver, signs of which animal were definitely present in the pit. If it really was the handiwork of *Eoanthropus*, the shaving could have been accomplished only with flint tools. Did flint knives have a sharp enough edge? Were they sturdy enough to slice through bone, and so delicately?

How had the slab, to start with, been removed from the dead elephant's thigh bone? Lopped off by percussion? By simple slicing or cutting? But the bones of dead elephants, it was known, when stripped of flesh weathered rapidly. They quickly became splintery, making them extremely hard to work.

If the implement—amusingly so like a small cricket bat—really was the work of the Piltdown individual, then it ranked as "by far the oldest undoubted work of man in bone." Supporting that contention was the clear trace of a small hole that had been bored through an edge up near the point. For one observer, this suggested that a thong had been threaded through the hole and bound tightly around the bone, making it serviceable as a club. Another thought it might have been used as a "hacking tool." In the end, on the question of use, it was agreed that the most sensible word had been spoken by a geologist who felt that it would be a mistake "to imagine too much specialization" in so primitive an instrument. Woodward's own final word, given at a regular meeting of the Geological Society that December, was unhesitating. The strange object was an authentic tool made by the no doubt still clumsy fingers of *Eoanthropus* wielding a flint knife. What this might reveal about the creature's mental abilities, social organization, or environment, he was content to leave for future consideration.

By now the Piltdown event—though interest in it had been curtailed and blunted by the outbreak of war in August 1914—had achieved worldwide stature. In all the paleontological circles of Europe and America the skull was a recognized object of study. At one point, inevitably, and largely because of the complexities that swarmed around it, there arose a brief flurry of rumors about its authenticity. Only once did the rumors reach print, however, and then only to be roundly denied. The brief reference came from an American geologist, William Gregory, who visited London on professional business in the summer of 1914.

"It has been suspected by some," wrote Gregory in an article in *The American Museum Journal* that same year, "that they [the Piltdown bones] are not old at all; that they may even represent a deliberate hoax, a Negro or Australian skull and an ape-jaw, artificially fossilized and 'planted' in the gravel-bed to fool the scientists." But no one believed such tales, he quickly added, which were thoroughly disproved by all the known circumstances of the discovery, stretching over a period of six or so years, and by the character of the men involved. "None of the experts who have scrutinized the specimens and the gravel pit and its surroundings," he summed up confidently, "has doubted the genuineness of the discovery."

The year 1915 became a culmination of sorts for what by now was called the Piltdown industry, with the term *Eoanthropus* installed permanently in the consciousness of evolutionary science. In February a long sketch of Woodward's professional career was included as a feature article in *Geological Magazine*. That same month an oil painting was commissioned for hanging in the Royal Institution, to show the men most closely linked with the discovery and its elaboration. The precious bones themselves were brought out of the vault at the Natural History Museum and put on public display in the museum's huge main hall, with a pamphlet by Woodward to explain them, *Guide to the Fossil Remains of Man*. That autumn, three significant books appeared in London, all written by leading men in the field, which described and interpreted Piltdown Man in minute detail.

In his *Diversions of a Naturalist*, Ray Lankester stated boldly that "in Eoanthropus we have in our hands, at last, the much-talked-of 'missing link,' " and he added that the Piltdown jaw was "the most startling and significant fossil bone that has ever been brought to light." Arthur Keith, despite his strongly divergent views on some things, in his comprehensive, five-hundred-page *Antiquity of Man*, almost half of which is given to Piltdown, stoutly echoed Lankester: beyond question, Piltdown represented "the earliest specimen of true humanity yet discovered." A similarly vigorous and positive judgment was offered in the widely read *Ancient Hunters*, written by the influential William Sollas.

Woodward's own planned book on Piltdown, to be called *The Earliest Englishman*, had gotten only so far as completion of the first chapter, setting forth the story of the discoveries. Ironically, it was Piltdown's fame, vastly increasing the burden of his professional duties, that had forced him to lay the book aside. Considering the intense public interest in the subject and in the gripping topic of evolution generally, it is a wonder that some publisher did not manage to coax a book from the one scientist who stood at the very center of the Piltdown phenomenon. Whether any publisher tried to do so, and failed, is not known.

Also attained in 1915 was a tidy simplification of the whole troubled Piltdown argument, when the embattled matter of the brain's size or capacity became secondary. (Woodward in his museum *Guide*, in any case, had quietly raised his own figure for cranial measurement closer to Keith's, making it 1,300 cubic centimeters.) Now the dispute became focused squarely on linkage of the jaw and the cranium. Did the two together legitimately form a complete skull? Or had the two fossils, one a man, the other an ape, somehow been swept into the same pit through an accident during the long centuries, propelled to each other's vicinity by shifting earth or rushing waters?

Responding to that tenacious query, a stream of informed commentary upholding one or the other position continued to appear on both sides of the Atlantic. Gaining most attention, and seemingly very troubling to Woodward and his friends, was one paper on the topic emanating from an unexpected source, a little-known anthropologist at the Smithsonian Institution in Washington, D.C., Gerrit S. Miller. His lengthy study, "The Jaw of the Piltdown Man" (thirty-one pages, well illustrated), published late in 1915, was based on comparative analysis of the actual jaws of more than a hundred chimps, orangutans, and gorillas. Its impact in paleontological circles, derived from its thorough coverage and controlled reasoning, was considerable.

Any attempt to combine the Piltdown cranium and jaw, said Miller, produced a freak, a weird primate that differed in its fundamentals from all other members of the order. The exact characteristics of a genus belonging to one family became superimposed on the characteristics of quite another family. Of such "contradictions," he specified, there were no less than four, all of which he spelled out in exhaustive detail: ball-and-socket mechanics (echoing Waterston), jaw musculature, spinal column derivation, and jaw-nosebone compatibility.

In each of the four instances, Miller added, these dramatically opposed characteristics were sharply defined in the fossils, and all were easily traced. Most telling, in no single feature was there any slightest sign of the blending of the two types, no adaptation of one to the other, which certainly could have been expected if an ape jaw had gradually accommodated itself to a human cranium.

Miller's sober and restrained conclusion was that by undeniable physical evidence, the Piltdown jaw and brain case must be seen as two separate entities, kept rigidly apart. This meant that each part or specimen became a valuable fossil in its own right: the brain case, though primitive, was fully human, the jaw was that of an ancient form of chimpanzee, dating from the Pleistocene. It was a type of chimpanzee new to science, so a proper name for it was required. *Pan vetus*, he aptly suggested.

Still, Miller conceded, two problems remained. First was the undoubted fact that the two specimens had been found closely associated in the ancient gravel. Statistically this circumstance undoubtedly favored linkage, but that was true only where no opposing factors existed. Where there is "obvious incompatibility," mere physical association "does not furnish serious elements of proof." The local soil conditions that brought one specimen to rest in the Barkham Manor beds "might be expected to act in about the same way with another." The second problem concerned the supposed absence of chimpanzee remains in Europe's Pleistocene, but on this there was little that could be said with certainty. Whether chimpanzees had in fact reached Europe so early might or might not be true. In Germany, at any rate, there did exist one claim for a Pleistocene chimp, though it was based shakily on a single tooth.

Woodward's response to the American's paper, expressed in a private letter, was unexpectedly harsh. Throwing off his aloof manner, in some heat he characterized Miller's article as "the latest ROT from the U.S.A.," adding that he was greatly surprised to find the respected Smithsonian deigning to publish "such nonsense."

In a more subdued key, most English paleontologists, though not all, similarly dismissed Miller's painstaking work. The main British reply was handled not by Woodward but by a colleague of his at the museum, the zoologist William Pycraft. His approach, made in public, took the form largely of personal abuse. (He would confess later that he had written while under a "feeling of irritation.")

The American, charged Pycraft in a twenty-page article in *Science Progress*, had made egregious errors in simple matters of observation. He had also "woefully misread" his many fossil specimens. He had shown a disgraceful "lack of perspective," and had obviously set out in the first place "to confirm a preconceived theory, a course of action which has unfortunately warped his judgment and sense of proportion." He himself, Pycraft said, had examined at least twice as many simian jaws and teeth as had Miller, and he had found none that could be at all compared with the Piltdown jaw to any purpose. The *Eoanthropus* skull, he stated peremptorily, despite all Miller's labored exposition, remained a perfect unit.

Maintaining his measured tone, Miller answered Pycraft with force some months later, producing another lengthy paper. Minutely he analyzed each of Pycraft's substantive points on tooth and jaw morphology, reviewed the history of scientific opinion on the topics, and added a few new arguments of his own. Until someone actually saw in a single animal skeleton the strange and inadmissible combination of a human cranium with an *unmodified* ape jaw, he insisted, the two parts must be attributed to different creatures.

It was an impressive performance by the unknown American. From his opponents in London, despite their disagreement, it must have brought at least a small portion of grudging admiration.

But by then, February 1917, it no longer mattered. A powerful new factor had unexpectedly entered the equation which would finally and dramatically tip the balance in favor of *Eoanthropus*. It made its appearance soon after the more or less sudden death of the fifty-two-year-old Charles Dawson.

4

The Sheffield Park Find

He was a talkative, busy, active man simultaneously pursuing what amounted to three demanding careers: in law, in amateur science, and in town government. With such constant, if willing, embroilment it is not very surprising that Charles Dawson should have found himself suffering from anemia. Consulting his doctor in November 1915 he was given that troubling diagnosis, and soon afterward he was put on a regimen of serum injections. Ignoring the doctor's advice to slow down, he continued to take charge of affairs at his law office in Uckfield. "But it was evident to all who saw him," reported one obituary, "that his health was impaired."

For a while in the early spring of 1916 he rested at home. Then, feeling somewhat recovered, he returned to work, but by summer he had once more been forced to keep to his house in Lewes, Castle Lodge. From that time "he was not seen in his familiar offices. He took to his bed and his condition became grave." By June an alarming condition of septicemia, or blood poisoning, had set in, the cause resisting diagnosis.

Now confined wholly to bed rest at his home, Dawson weakened through the heats of July. Late that month he was cheered by a visit from Woodward and Smith, who were spending two full weeks as guests at nearby Barkham Manor while they continued the digging (nothing new had turned up). Some improvement through the early part of August gave brief hope that he would recover. But then during several days he grew steadily and rapidly worse, and early on the morning of August 10 he died. The death certificate stated that his bout with septicemia had lasted for forty-two days, and that the anemia had begun at least nine months before. Also, for five months the dead man had suffered from "pyorrhea alveolaris," ulceration of the gums.

Several weeks later, Dawson's grieving widow, in writing to an old friend, fondly and sadly recalled her husband as "the best and kindest man who ever lived, and you, who knew him so well, could not fail to love him." It was a description faithfully echoed in the death notices and in the spontaneous reaction of his many acquaintances.

At a regular weekly session of the Uckfield Magistrates Court—Dawson had been its chief clerk for twenty years—one of the judges delivered an impromptu eulogy. Recalling "that quiet, unassuming man, with pleasantly smiling eyes gleaming through rather obtruding spectacles," he emphasized Dawson's high professional integrity. It could be seen, he said, in his meticulous case preparation and in his always fair-minded handling of the evidence: "There was one thing about Mr Dawson; he was always sure. He never gave an opinion, to his knowledge, unless he was sure, he was never ashamed to look it up. When he was a practicing barrister he had a great deal of experience of magistrate's clerks. They were of all sorts and types. Many of them did all they could to secure convictions, but Mr Dawson never did anything of the kind." The eulogist also referred to Dawson's "brilliance" as a geologist-antiquarian and drew a contrast between the man of so many accomplishments and his rather ordinary air and appearance: "If anyone had seen [him] issuing from the door of a matter-of-fact country lawyer's office they would not have realized what a man of romance he was."

Both Woodward and Keith wrote lengthy notices extolling the dead man's scientific accomplishments, Woodward in *Geological Magazine* and Keith in the *British Medical Journal*. For Woodward, he had been "a genial presence . . . a delightful colleague in scientific research, always cheerful, hopeful, and overflowing with enthusiasm." Keith, whose personal contact with Dawson had been minimal, and mostly through Woodward, fittingly praised him as "a splendid type of that great class of men who give the driving power to British science—the thinking, observant amateur. Without that class to serve as scientific scouts we could make little progress in our knowledge of the past."

It was *The Sussex Daily News* that put on record the early start that the "observant amateur" had made in his scientific career, stating that "as a boy of twelve Mr Dawson was already beginning to form a collection" of fossils. The precocious youth, added the paper, was so preoccupied with geology that he "spent all his pocket money on buying fossils from the quarrymen at Hastings," and passed nearly all his playtime hours in "tracing the footprints of giant fossil reptiles in the Wealden rocks, digging out the monsters' bones and piecing them together." To his friends and neighbors, who looked on Dawson's activities as a harmless pastime, it had come as something of a shock to see experts from the British Museum arrive at the Dawson home in 1884, when the young man was only twenty, "spending whole days in carefully packing the specimens for their safe journey to London." The *Daily News* also preserved the colorful title bestowed on Dawson locally, a name that amused but also greatly pleased him: "The Wizard of Sussex."

The funeral, held in Lewes on Saturday, August 12, 1916, was attended by a large crowd of mourners, probably in excess of a hundred, many arriving from London, his wife's former home, and from more distant places. *The Sussex Daily News* listed fifty-six names as "among those present," aside from family and friends, and also named the contributors of the more than forty floral offerings that filled the church. On hand were Dawson's two brothers and his sister with their families. One brother, Reverend Leyland Dawson, vicar of Clandown (near Bath), helped conduct the services. "It was interesting," said the *Daily News*, "to note among the general body of mourners in the congregation Dr. Arthur Smith Woodward of the British Museum," beside him in the pew his son Cyril.

Woodward's feeling of sorrow at his friend's passing, though the two had never been intimates, was genuine. But he also had another, more immediate, even an urgent reason for concern. Involved was yet another discovery of Dawson's, made no less than a year before but which had not yet been revealed. No common or ordinary discovery, it supplied physical proof of the existence of a second Piltdown Man, a development that would go far toward establishing among scientists the complete authenticity of *Eoanthropus*. Announced suddenly and for the first time to the world by Woodward in February 1917, its prior history has always been curiously problematic.

To the extent that the facts are now traceable, it can be said that Dawson's last discovery took place during the first half of 1915, and in three stages. Unexpectedly, it was not associated with the Piltdown site, but with a second locale, an open field bordering the River Ouse some two miles west of Barkham Manor. The precise location of this field was never stated, though Woodward later privately reported it to be in or adjacent to the hamlet of Sheffield Park.

Through 1914, the relentless Dawson had been gradually extending his search outside the Piltdown area, tracking the spread of the ancient gravels. As he made each of the new finds, starting with a small piece of cranial bone, identified as from the forehead, he duly reported it to Woodward. "I believe we are in luck again!" he announced happily in a note of January 9, 1915. "I have got a fragment of the left side of a frontal bone with a portion of the orbit and root of the nose." The new bone, he added, closely resembled Woodward's original restoration, "and being another individual the difference is very slight . . . the general thickness seems to me to correspond to the right parietal of Eoanthropus . . . the forehead is quite angelic!" Then, some weeks later, he reported picking up in the same field a small section from the back of the skull, a bit of the occiput.

In July, Dawson wrote a third time to say he had come across a tooth, which he reported to Woodward not by letter but in a postcard. The excitement of discovery seems by now to have been somewhat dissipated, for he mentions the highly significant new find only after covering several other topics of casual interest:

I am sorry I shall not be able to come to Keith's demonstration. I wonder if he has seen the new paper on Pleistocene and Pliocene Primates in the Indian Gov. Geol. Publication. The writer sent me a separate but I forget his name.

Boule has been writing me about Starch Eoliths and I have sent him a sample of them. He has promised to send me a separate of some new paper he has written about Eoanthropus.

Teilhard wrote yesterday—he is quite well and in a quiet spot at present. I have got a new molar tooth (Eoanthropus) with the new series. But it is just the same as the others as to wear. It is a first or second right m. The roots broken. At Piltdown the road is falling in at the edge where we got too close and Kenward has put up a fence.

For whatever reason, during Dawson's remaining year of life no public reference was made to these new specimens, no report, formal or informal, was made to any scientific body. Nor in their correspondence of 1915–16 did the two men have anything to say about them (for the six months prior to Dawson's death no letters between the two have so far surfaced). Dawson's illness, starting in the fall of 1915 and slowly growing worse through the first half of the next year, may have been some part of the reason for the delay. But it would not account for the three months of inaction just after the finding of the tooth, and would not explain why Woodward might not have proceeded on his own with so important a development. Perhaps the two men were holding back in hopes of finding yet further specimens at the new location.

In any case, the existence of the Sheffield Park bones was not kept entirely a secret, for in the summer of 1915 Dawson showed bits of the cranium to at least one other qualified scientist, the eminent zoologist Ray Lankester. In his book, *Diversions of a Naturalist*, just then going to press, Lankester managed to insert a footnote to his long and favorable discussion of Piltdown. Briefly and rather cryptically he noted "the recent discovery by Mr. Dawson of a second skull of the same character as the first." Tantalizing as that footnote must have been to readers on publication of Lankester's book that September, it passed without remark. (The books of Keith and Sollas, both published that fall, have no mention of a second skull.)

Both pieces of the Sheffield Park cranium, along with the tooth, had remained in Dawson's hands, kept safely at his home in Lewes. On his death, the three pieces were duly passed to Woodward by the widow, at Woodward's own request, as it appears. But now another unexplained delay occurred, five more months passing before Woodward felt ready to make a formal presentation at a meeting of the Geological Society. This time, however, the reason for the delay becomes obvious: nothing else than Woodward's embarrassment at being unable to say just how, and especially just where, the finds had been made.

As Woodward would later explain, Dawson "told me that he found them on the Sheffield Park Estate, but he would not tell me the exact place—I can only infer from other information that I have." So far as it can be interpreted, that other information located the site on land near the village of Little Chailey, just bordering Sheffield Park, possibly a farm with the name Netherall. But beyond that Woodward could not go. Anxiously he questioned Dawson's widow and the clerks in Dawson's Uckfield office, among others, but to no avail. Woodward's wife later vividly recalled this uneasy and troubled interlude, explaining that "Dawson would not give details of the exact spot," and that her husband on his own "spent much time searching for it" fruitlessly. When Woodward finally gave his presentation to the Geological Society in London, on February 28, 1917, facing a large and attentive audience (smaller than it might have been because of the war), he could not resist fudging the facts more than a little:

> One large field, about 2 miles from the Piltdown pit, had especially attracted Mr Dawson's attention, and he and I examined it several times without success during the spring and autumn of 1914. When, however, in the course of farming, the stones had been raked off the ground and brought together into heaps, Mr Dawson was able to search the material more satisfactorily; and early in 1915 he was so fortunate as to find here two well fossilized pieces of human skull and a molar tooth, which he immediately recognized as belonging to at least one more individual of *Eoanthropus dawsoni*.

Strongly Woodward implied, if he did not distinctly state, that the precise location of Dawson's find, and its circumstances, were all well known to him, as would only be proper. What he had done, apparently, was to conclude or assume that the finds were made in some otherwise unidentified "large field" to which Dawson had led him in 1914. That he omitted naming the locality in his talk would not have disturbed his listeners, for it was the only way to ward off destructive curiosity seekers. Evidently his quiet deception succeeded nicely, for in the discussion that followed the meeting no one ventured to raise the question of location, ordinarily of some real interest.

A corroboratory animal tooth, Woodward also announced, had been found in the same unnamed field and in "the same gravel." It proved to be part of the lower molar of an ancient rhinoceros, "of indeterminable species," and was as highly fossilized as the corresponding animal teeth from Piltdown. Its finder, he said, was not Dawson but "a friend" of Dawson's, also unnamed. Before the meeting, Woodward had tried hard to find who this friend might be, again questioning Dawson's Uckfield contacts, but had again failed, a fact he did not bother to state to the audience at the Geological Society. Nor did the friend ever come forward, then or later, and his identity remains unknown.

The audience discussion was brief and, this time, free of contention. Only three speakers rose to be heard—William Pycraft, Arthur Keith, and Ray Lankester—and all three expressly agreed with Woodward's unflinching summation. From his close study of the new specimens, Woodward had said, it seemed reasonable to conclude that *Eoanthropus* would prove to be as definite a form of early man, and every bit as distinct, as he had at first supposed. "The occurrence of the same type of frontal bone," he stated pointedly, "with the same type of lower molar in two separate localities adds to the probability that they belonged to one and the same species."

In that professional audience, all knew well what Woodward meant by the quiet mention of "probability" in connection with the "two localities." Perhaps it might happen once, accidentally and randomly, that a real ape's jaw should wind up in the same pit alongside a human brain case, both of them fossilized. That sort of accident was admissible, though barely. But the odds against such a chance pairing of opposites, of unrelated but similar artifacts, occurring a second time in the identical vicinity could be said to verge on the mathematically impossible. It was clear that the Sheffield Park finds established the incontestable reality of Piltdown.

Gradually, over the next several months and years, accelerating with the war's end in the fall of 1918, many experts who had vehemently opposed the combination of jaw and cranium now simply gave way, confessing themselves wrong. The conversion, while by no means complete, was dramatic and sweeping, affecting some of the leading names in the field worldwide—for instance, one of the best-known, Marcellin Boule of France. The reaction of one American anthropologist, long an opponent of linkage, wonderfully illustrates the transposing impact of the new discovery.

In the summer of 1921, Henry Fairfield Osborn, president of the American Museum of Natural History, visited London. It was a professional visit, so he gladly accepted Woodward's prompt invitation to make a personal inspection of the Piltdown bones, new and old. At the museum on a Sunday morning in July with his host he was conducted in some ceremony to a special room in the basement. There a steel safe was opened and the "few precious fragments" were taken out and arranged on a table. For two uninterrupted hours the skeptical Osborn peered and measured, his comparisons patient and painstaking. Suddenly he found he could no longer hold out.

"If there is a Providence hanging over the affairs of prehistoric men," he said afterward, "it certainly manifested itself in this case." The three small fragments of the second Piltdown man, he marveled, "were exactly those which we should have selected to confirm the comparison with the original type." Just those specimens which were most needed by the advocates of *Eoanthropus* had actually been found. Inspecting the second Piltdown cache, placed side by side with the corresponding fossils of the first, Osborn was forced to admit that "they agree precisely; there is not a shadow of difference."

To the hovering Woodward the contrite American made his confession. The chinless Piltdown jaw, so like a chimpanzee's, and the skull with its high human forehead and capacious brain case, did indeed and beyond question belong together. Confounding all the meticulous, too-careful experts, Piltdown Man was a triumphant reality.

On the following Tuesday, Woodward had a treat for his guest. Together the two men boarded a train in London and traveled happily down to Uckfield, then motored over to the Piltdown pit to do some token digging. With them went a photographer, and his picture of the moment captures the two scientists standing side by side at the pit's edge. Straight into the camera they stare, sober-faced and satisfied. Their jackets are off but their vests are neatly buttoned over white shirts, and their ties are in place. From Woodward's right hand casually dangles a small geology hammer.

With announcement of the Sheffield Park finds the story of the Piltdown discoveries comes to an end. Despite the many hundreds of hours spent by Woodward and his helpers digging at Barkham Manor in the ensuing twenty years, despite a final spurt of digging in 1950 just before the site was designated a national monument, not another fossil of any kind was turned up.

But hovering quietly in the background all this while was still another small cache of fossils found and preserved by Dawson. Now known as the Barcombe Mills finds, in the original Piltdown drama they had no role. Ever since they have been largely and strangely ignored. Yet in a peculiar manner they do have a place in the story, even a substantial one, which will receive its explanation in the unraveling further on. The record of their provenance, however, belongs here, and it is quickly told.

Barcombe Mills is a village in Sussex nestled against the River Ouse just north of Lewes. Its first mention in the Piltdown documents occurs in a letter from Dawson to Woodward of November 1912. Among other matters he reports that with some geological friends he has been out prospecting for likely fossil sites, has visited some fields beside the river at Barcombe Mills, and has there picked up several ancient flint implements. Seven months after that, on July 3, 1913, Dawson writes again to inform Woodward of a possibly important discovery at Barcombe Mills:

> I have picked up the frontal part of a human skull this evening on a ploughed field covered with flint gravel. It [is] a new place, a long way from Piltdown, and the gravel lies 50 feet below [the] level of Piltdown, and about 40 to 50 feet above the present river Ouse. It is *not* a *thick* skull but it may be a descendent of *Eoanthropus*. The brow ridge is slight at the edge, but full and prominent over the nose.
>
> It was coming on dark and raining when I left the place but I have marked the spot. The base of the nose is very rotten so I have put some of Barlow's mixture [on it]. Will you get Barlow to give us the recipe for gelatinizing, as the bone looks in a bad way and may go wrong in drying. I have got a saucepan and gas stove at Uckfield . . .

Missing entirely from the record is any response to this intriguing news on Woodward's part, and during the next four years, while Dawson lived, nothing at all is heard of Barcombe Mills. Following Dawson's death in the summer of 1916, his widow set aside for Woodward's personal inspection all the remaining fossil material she was able to locate at Castle Lodge. Exactly when Woodward's visit to Lewes took place is not known, but in January 1917 he made entries in the formal register of the Natural History Museum regarding not one but four specimens identified as coming from Barcombe Mills. Included were two pieces of human cranium (frontal and parietal), a pair of small zygomatic bones (cheek), and a single tooth, a lower right second molar.

Accompanying the two cranial fragments is a notation which states that they were found in "Pleistocene gravel in field on top of hill above Barcombe Mills railway station." Concerning the molar, another notation reads, "probably from the same place (not certain)." Woodward never made any reference, public or private, to these four items, and no additional information survives about their origin and handling. As with the Sheffield Park incident, their full story, if it was ever known, has been lost.

For thirty years the Barcombe Mills materials lay forgotten in a storage room at the museum. Then in 1949 they were found, identified, and closely studied, upon which their peculiar worth was at last recognized. "Smith Woodward probably considered the Barcombe Mills skull of no particular scientific value," reads an early report on them by Robert Broom, the well-known South African paleontologist, "and he made no reference to it in his little book 'The Earliest Englishman' published in 1948 after his death. As it appears an antique skull and from Piltdown gravels, and thus a very early type of human skull, it seems worthy of rather intensive study in connection with the two previously known Piltdown skulls . . ."

The decision rendered on the Barcombe Mills bones was not undisputed, being opposed on various counts and in varying degrees of heat (one expert saw them as being from several individuals of the Neanderthal type). But there were at the same time many paleontologists who readily and happily agreed with the conclusion that in the long-neglected Barcombe Mills specimens there resided yet a third *Eoanthropus.*

❈ 5 ❈

Curtain Rising

The large banquet room of the Rembrandt Hotel, across Cromwell Road from the Natural History Museum, was crowded with scientists and their guests. All were on hand to attend an international conference at the museum on early man in Africa. At one table along with several other guests sat Joseph Weiner, thirty-eight-year-old professor of physical anthropology at Oxford. Seated beside him was Kenneth Oakley, well-known paleontologist and a member of the museum's staff. It was Oakley who nearly a decade before had devised the fluorine tests that so drastically reduced the evolutionary importance of Piltdown Man. The two principal performers for the final act of the lengthy drama were on stage.

Earlier that day in the museum's geology department, Weiner along with many others had gotten his first fascinated look at the original Piltdown remains. During almost four years he had been nursing an intense curiosity over the troubled picture of Piltdown Man that had been established by Oakley's initial report, turning the creature into hardly more than a grotesque survival, an evolutionary dead end. The sight of the original bones had sparked in Weiner all the old nagging questions, and the talk at the table between the two men soon centered on Piltdown. Eventually the conversation narrowed to the Sheffield Park finds, which by now had become known as Piltdown Two.

In company with other anthropologists, Weiner had always assumed that the exact site of the Piltdown Two discoveries was known to those concerned, and that knowledge of the site had been kept from the public as a safeguard. Now, for the first time, he found himself wondering why so important and promising a site had never been fully excavated, or so he understood. Asking Oakley about the omission, he was informed, almost casually, that the museum had no record of the precise location of Piltdown Two. "The fact is," admitted Oakley, "all we know about site II is on a postcard sent in July 1915 to Woodward, and an earlier letter in that year, from neither of which can one identify the position of Piltdown II."

Here, thought Weiner, was a decidedly "curious piece of information." Piltdown Two had been the critical factor, the turning point, in gaining acceptance for the main Piltdown finds. Yet now, it turned out, almost nothing was known about the circumstances of the discovery! Why hadn't Dawson recorded at least the essential facts before he died? Had Woodward in reality been told about the site, and had he then mislaid his notes? Why had nothing at all ever appeared in the literature of paleontology about so remarkable an oversight?

Stirred by the unexpected disclosure, Weiner drove home to Oxford that night—it was July 30, 1953—with his restless mind churning. He reached his apartment after midnight, but found that he couldn't get to sleep with thinking of one worrying question: if the Sheffield Park bones were in any way questionable, how did that affect the original Barkham Manor finds? Immediately there pressed into view the old nagging doubts about the jaw-cranium linkage.

Either the jaw and the cranium came from a single individual, a veritable man-ape, or they came from two different creatures, a primitive man and an early ape. Weiner himself had always been inclined to the latter view, opposing linkage, and now for the hundredth time he asked himself the pivotal question. *How* had all those bones come together in the ancient soil at Barkham Manor? By accident? Had someone, years before, casually dropped an ape jaw into the pit, almost on top of the cranium? But even had such an unlikely event taken place, surely it could never have been repeated at Sheffield Park?

Perhaps the true significance of site Two had been exaggerated, somehow wrongly interpreted. The total absence of knowledge as to its precise location—the most fundamental sort of information in the science of paleontology—was enough in itself to raise doubts. Then another possibility dawned: could the bones from Piltdown Two have actually been a part of the first find, the original discoveries at Barkham Manor? Had they been carted away from the manor, mixed in with the piles of rough flints being used for road mending in the Sheffield Park vicinity?

But all such groping speculation, Weiner suddenly realized, entirely avoided the main issue. Wherever the Piltdown jaw came from, however it had reached the Barkham Manor pit, it was still a fossil jaw. Two unrelated bones, both fossils, both supposedly dating from the early or mid-Pleistocene, coming together in the ground by *accident*, perhaps even an accident of human agency? Here Weiner's whirling thoughts, as he recalled, began to pull free from the drag of long-accepted belief.

Say that Oakley's fluorine tests, he reasoned, were in some way flawed or inadequate—they did, after all, carry a hefty plus/minus factor. In that case the ape jaw might not be very old at all. It might be modern. But there were the teeth, the two molars and the canine. The pattern of wear in them was definitely *not* that of an ape. It was human. Here came the final mental leap to the theory of fraud, a moment that was described by Weiner soon after: "A modern ape jaw with flat worn molars and uniquely worn-down eye-tooth? That would mean only one thing: *deliberately* ground-down teeth. Immediately this summoned up a devastating corollary, the equally deliberate placing of the jaw in the pit."

If the unusual teeth were spurious, produced by artificial abrasion, reasoned Weiner, then everything connected with the jaw came into question. Missing were the symphysial region (chin), as well as the condyle (swivel knob). Both of these absences strongly supported the idea of a planned and deliberate imposture, for they were precisely the features a hoaxer would eliminate. Now the jaw's being in the pit at all became doubly suspect. The mysterious hand that had so cleverly altered its color and configuration must also have placed it where it was found, or arranged for the planting.

Next morning early, the anxious Weiner was at work in his laboratory, minutely inspecting the Piltdown casts. Nothing he saw in the next few hours made him hesitate over the theory of fraud. All the suggestions that had been made regarding the variety of "human features" to be traced in the jaw's morphology, Weiner now realized, could be reduced to a single one, the teeth, the two molars and the canine. Everything else supposedly human about the jaw could be discarded as unproved and unlikely. How curious, he thought, that only a single feature, the teeth, should provide the link between jaw and cranium. Why were there not "a few other modifications" by way of adaptation, definite changes that would support the testimony of the teeth? It was at this juncture that he recalled the old objections to the canine, the apparent contradiction between its age—only recently erupted—and its worn condition.

Inspecting the cast of the jaw through a lens, Weiner noted several things that had never impressed him before, and which he could not recall hearing or reading about. For instance, the edges around the flat biting surfaces of the molars were all quite sharp, not beveled or softly rounded from use, as might have been expected. Also, the two biting surfaces themselves were not quite in the same horizontal plane, were not continuous, but pitched slightly opposite to each other. While the significance of such a misalignment was uncertain, it provided another anomaly to be noted. A third suspicious factor about the teeth he now saw that had escaped comment: the degree of wear in the two molars was exactly the same, almost no difference. But an inside tooth, it was known, nearly always wore down more speedily than the one farther out.

Turning to Oakley's 1949 paper on the fluorine tests, Weiner abruptly spotted something that he, and everyone else, had missed on previous readings. When drilling for a sample of dentine, Oakley had found that just beneath the dark, iron-stained surface, the dentine was pure white, "apparently no more altered than the dentine of a recent tooth from the soil." Such a superficial stain, it struck Weiner, argued recent application: the forger would of course give the teeth the appearance of age by staining them the approximate color of the soil in which they were to be found. But a stain that had supposedly soaked in from the iron-bearing soil over untold centuries should have penetrated much deeper than the thin surface layer described by Oakley.

A final disturbing fact showed up as Weiner perused the original paper given by Dawson and Woodward in 1912. The cranium, explained the old, formal report, had been tested for its organic content—revealing the degree of fossilization—and for other conditions by Dawson's friend Samuel Woodhead. But nothing was said about similar tests being made on the jaw. Here was a neglect of proper procedure for which no explanation was given, and for which Weiner, forty years later, could imagine none.

His suspicions now fully aroused, Weiner made the logical decision to try producing a fake tooth of his own, one to match those in the jaw. From Oxford's anatomy lab he procured a modern ape's molar and carefully filed the biting surface flat. Even with rough tools and no experience, he found, the job proved surprisingly simple. Using permanganate, he then stained the tooth's whole surface a dark soil color, and the result was a fossil tooth looking very much like the molars in the Piltdown jaw, though lacking their finish.

He was dealing, Weiner knew, with an issue of utmost sensitivity, one that could inflict great damage on the reputations of many scientists, dead and alive. Wishing to consult privately with a knowledgeable peer, a week after beginning his investigations he quietly reported his results to his department head at Oxford, anthropologist Wilfrid Le Gros Clark, a man long familiar with all aspects of the Piltdown story. Listening patiently to Weiner's earnest explanations as they both peered closely at the filed-down teeth, Clark was surprised to see how exactly they mirrored the unusual type of wear in the Piltdown molars. At last he exclaimed, "You don't really mean to say that is the way it was done!" When Weiner responded with a decided "Yes!" Clark threw aside all hesitation. "I am sure you are right," he stated enthusiastically as he saw how the thesis of forgery at Piltdown would instantly clarify the still much agitated evolutionary picture.

Oakley, he suggested, must be informed immediately, but nothing should be committed to paper, officially, at that early stage. Instead, they would telephone directly to the museum in London.

It was late on the afternoon of August 6, 1953, that Kenneth Oakley in his office at the Natural History Museum picked up the telephone to hear the voice of his friend Le Gros Clark calling from Oxford to tell him that the Piltdown jaw was almost certainly a forgery. Piltdown Man as it had been known during four decades, stated Clark succinctly, did not exist. Later Weiner recalled Oakley's reaction to the devastating news: he was "quite taken aback."

Most urgently, Clark and Weiner wished to have Oakley take one simple step immediately. Using the strongest magnifier available, the two molars and the canine must be inspected to determine if any sign was present of artificial abrasion. The agitated Oakley said he would proceed without delay to make the inspection and would phone back as soon as possible. An hour passed, and then Oakley was on the phone again. He was "utterly convinced that abrasion had been applied, and it was particularly obvious on the canine." Within days, plans were laid for full-scale testing of the whole jaw, with the museum itself assuming the lead.

All the men involved had other urgent duties, so the investigation was slow in starting, but in a bit more than two months the verdict was reached. It left no doubt that the Piltdown jaw was indeed a forgery, a cleverly doctored chimpanzee jaw or perhaps that of an orangutan. Also eliminated were the bone fragments from Piltdown Two—Sheffield Park—not only the molar tooth but the two cranial pieces as well.

Four principal areas were covered by the tests, starting with abrasion of the teeth. Weiner's own surmises about the evidence of the teeth—the unusually sharp edges, the variety of wear in the molars, the out-of-line surfaces—were all confirmed. In addition, several other factors were uncovered, more minute and technical but equally revealing. In the second molar, for instance, the center of the "talonid basin" was found to be completely unworn, yet its surrounding edges were unnaturally thin, a condition that could not have resulted from ordinary wear. The exposed dentine at several points also spoke of artifice, particularly for being more in evidence at some places than others, most of them demonstrating just the opposite of normal attrition.

The wear pattern on the canine, now proved to be a young tooth, was glaringly artificial, "unlike that found normally either in ape or human canines." The molar tooth from Piltdown Two, looked at under a binocular microscope, revealed that "the occlusal surface of the enamel has been finely scratched, as though by an abrasive." The pronounced flatness of the biting surfaces, with their unbeveled margins, was the result of deliberate grinding with a metal tool.

The second and third areas of study involved chemical tests, one of them a new fluorine measurement (the technique had been much improved), the other a probe of organic content focusing on nitrogen loss. The two tests were designed to complement each other, and in this instance they clearly agreed: the Piltdown jaw and canine, and the Sheffield Park molar, were undoubtedly modern.

The final revelation came with the staining, or coloring, of the specimens. The blackish outer coating of the canine, early judged to be an iron compound absorbed from the soil, was now revealed as nonmetallic. It actually consisted of a "tough, flexible paint-like substance," not then identifiable (later it was declared to be an actual paint, based probably on ordinary Vandyke brown).

In many of the bones, chromate was found to be present, but not all, and this posed a minor mystery. The pieces Dawson had recovered before the advent of Woodward had all been treated with potassium bichromate to harden them, as Dawson explained. Woodward had asked that the procedure be stopped, and the pieces found thereafter bore no trace of chromate—all but one, the jaw. Chemical treatment of the jaw could not have been done after the discovery—at which Woodward was present—so it must have been done before, and the purpose was transparently to make it more closely match the cranium in color. But here the forger stumbled somewhat, for the staining on the skullcap penetrated deeply into the bone, where the jaw's stain remained superficial.

The culminating revelation in these tests came with the finding that the cranial pieces from Piltdown Two also contained chromate, thus had been similarly treated and for the same reason. In addition, it was now concluded on the basis of further chemical testing that the small frontal fragment from Piltdown Two—the piece of forehead, as well as the isolated molar—had all originally been part of the first skull found at Barkham Manor. The supposed second Piltdown Man did not exist either.

As the astonished investigators admitted afterward, the distinguished Piltdown scientists of forty years before had been the unwitting victims of "a most elaborate and carefully prepared hoax." The faking of mandible and canine was "so extraordinarily skilfull, and the perpetration of the hoax appears to have been so entirely unscrupulous and inexplicable, as to find no parallel in the history of palaeontological discovery."

At no time among the investigators or the staff of the museum was there any thought of withholding or disguising any part of the outcome. By late October, when the results had become inarguable, plans were made for a prompt release of everything to the press. Some little time was consumed in preparing the official paper, minutely describing and explaining what had been done, but by mid-November all was ready. On the twentieth, copies of an illustrated museum bulletin entitled *The Solution of the Piltdown Problem* were distributed, and the next day the *Times* headlined, PILTDOWN MAN FORGERY / JAW AND TOOTH OF MODERN APE / "ELABORATE HOAX." Quoting from the bulletin, the *Times* told with satisfying accuracy and detail the full story of the "startling discovery."

Less restrained, the London *Star* announced THE BIGGEST SCIENTIFIC HOAX OF THE CENTURY. In a weekly, *The People*, the headline predictably blared, GREAT MISSING LINK HOAX ROCKS SCIENTISTS. A few days later an editorial in the *Times* reassured more informed readers that "it is opportune to consider how the discovery of this curious imposition affects the general present understanding of the evolution of man." A lengthy column of discussion followed, enumerating the legitimate fossils and their significance, and ended, "Much remains to be learnt. The solution of the Piltdown problem is an important step forward in understanding."

That same afternoon, Weiner and Oakley left London by auto for the village of Downe in Kent, where they called by appointment on the eighty-six-year-old Sir Arthur Keith. In the years since the original Piltdown controversy Keith had constantly returned to the topic, never quite satisfied that he had exhausted its full evolutionary meaning. In his 1938 book, *New Discoveries Relating to the Antiquity of Man*, he had revived active discussion of Piltdown by bravely fitting it into the now greatly expanded evolutionary picture. The year after that, he had performed a full-scale resurvey of the Piltdown remains, producing a twenty-page article for *The Journal of Anatomy*. In his autobiography, completed and published just after World War Two, he provided a nostalgic, six-page account of his own role in the Piltdown story. In it he conceded that there were many questions about Piltdown still to be answered, but he also continued to defend its high importance in the fossil record. In the foreword he wrote for Woodward's 1948 book, *The Earliest Englishman*, he staunchly declared that "no theory of human evolution can be regarded as satisfactory unless the revelations of Piltdown are taken into account."

It was to soften the blow they knew Keith would suffer from the sudden extinguishing of his old enthusiasm that Weiner and Oakley had traveled down to Kent, intending to lend personal sympathy and understanding. As it turned out, they were just too late, for Keith had seen the *Times* story early that morning, and its effect on him was apparent.

To Weiner, shaking his hand in greeting, he "did not look very well, he was pale and rather pinched looking and shaky and had a cough."

In the elderly anatomist's study, surrounded by a half-century's accumulation of books and papers, the little group talked, Keith all the while seeming anxious for understanding "and indeed for sympathy . . . He was certainly rather bewildered with it all." Demonstrating their explanations on a cast of the jaw, the two guests offered a quick, basic review of the scientific aspects of the forgery and were pleased to find that Keith seemed well able to grasp what was said (though he kept forgetting Woodward's name and had to be reminded). It was a greatly saddened Keith who finally commented, "You may be right and I must accept it. But it will take me a little while to adjust to it."

The meeting lasted more than an hour, the talk wandering from science to personal matters, with Keith several times exclaiming in wonder how he could have been so completely fooled, and more than once attempting to justify himself (he *had* doubted the jaw at first, but there was the flat wear in the molars, and the cranium had "orang-like features"). It was the finding of the canine, above everything, that had convinced him, he volunteered, the canine with its peculiar pattern of wear so unlike anything ever seen. When Weiner ventured to ask whether he had ever considered that the canine *wear* might be artificial, "he did not reply." Almost everything the trio discussed that afternoon—the truth about the paint stain on the canine, the superficial staining of some of the bones, the misplaced dentine exposure, the altered planes of the two molars—all greatly "surprised" and "astonished" Keith to hear.

Had he never been disturbed or puzzled, asked Weiner, by the way the various finds were made in neat succession, each one offering timely support for the initial discovery? "Yes!" replied Keith, suddenly aroused, "I almost felt that they kept on finding things with which to confute me personally!"

In announcing the forgery, Weiner and his colleagues had, without stressing the fact, also explained that the Piltdown cranium itself, the skullcap, was still to be regarded as a true fossil. In testing the cranium for fluorine and nitrogen, nothing had surfaced to cast doubt on its authenticity—information which the *Times* story had duly conveyed to the public. "The fragments of the cranium are genuine remnants of primitive man," it stated, adding that their age could be placed at about fifty thousand years, agreeing with the results of the first fluorine tests of five years before. Weiner and Oakley soon went further, suggesting that the hoax had been built on an actual discovery at Piltdown of fragments of a true fossil cranium.

It was Kenneth Oakley, so far as is known, who first changed his mind, concluding or surmising that the cranium too might be a fake, and calling for full-scale testing to continue. With that, a massive new effort was quickly mounted, taking in the Piltdown material from all three sites, every last item down to the least impressive flint eolith, and drawing on the skills of a dozen scientists in several laboratories. Before the arrival of summer 1954, the verdict was at hand: *everything* was fraudulent. Everything found in the Barkham Manor gravel beds, as well as everything retrieved from Sheffield Park and Barcombe Mills, was either a forgery or an authentic piece deliberately planted. None of the total of more than forty specimens (counting the questioned eoliths) from the three sites had occurred naturally. The Piltdown hoax, it appeared, was infinitely more "elaborate" than anyone had guessed or could imagine.

Where the cranium was concerned, the first and most decisive indication of fakery showed up in a crystallographic analysis, which detected gypsum in the bone. As tests proved, this element could not have come from the Piltdown soil, and it was soon demonstrated that a fossil bone soaked in iron sulfate undergoes a chemical alteration leaving gypsum as a by-product. The bogus iron-staining of the cranium, it was decided, had been obtained in just this manner. Also detected in the skullcap was another alien substance, chromium, explicable only as the result of soaking in a dichromate solution to aid oxidation of the iron salts.

The telltale gypsum also showed up in the slivers of turbinal bone (nose), showing that they too had been artificially stained. For the two cranial pieces from Sheffield Park the same held true. Other tests—for instance to measure carbon content and to gauge radioactivity—confirmed all these findings.

The great thickness of the skull wall, held by the original discoverers to be an indisputable sign of prehistoric origins, was now shown to be explicable on other grounds. Thickness in modern skulls, it was determined, though rare, did in fact occur. One such skull was unearthed in the cluttered basement of the Natural History Museum itself, where it had been stored forgotten for many years, that of an Indian from Tierra del Fuego. But if the thickness did not occur naturally, then it could be explained as the result of disease. As had been known from the first, the Piltdown skull wall's marked increase in width was all in the *diploe*, the honeycomb of brittle bone sandwiched between the solid upper and lower plates, both of which remained very thin. This inner portion of the skull wall, it was agreed, might very well have been affected by a pathological condition, perhaps severe chronic anemia. At least one of the investigating team believed firmly that the thickening was "incontestably the residue of a morbid process." Support for that conclusion was found in studying a series of authentic fossil skulls. In every instance, the thickening was as pronounced in the plates as in the *diploe*.

Unexpectedly, an interesting and revealing, if tangential, result showed up as the specimens were being readied for testing. When drilling into the jaw to obtain small samples of the bone, the operator detected a strong smell of burning, and he noted that the ejected material consisted of minute shavings. In drilling the cranium, on the other hand, there was no odor and the sample came out as powder. Here was indicated a great disparity in organic content, thus in degree of fossilization, further separating jaw and cranium.

The teeth also gave up much valuable new information. The canine's inner chamber, or cavity, was found to be large and open, another definite sign of immaturity, again contradicting the tooth's much-worn surfaces. In the cavity were packed tiny granules of sand. When the diameters of these grains were measured, all were found to be much alike. But a natural infiltration from a riverbed would have introduced grains of different sizes. On no tooth could there be traced any sign of secondary dentine, a natural replacement readily to be expected in an older individual. Nor did the teeth reveal any evidence of postmortem damage, probable in a prehistoric tooth that had been tumbled about in disturbed soils and moving river currents. The roots of the two molars in the jawbone, apparently quite human when viewed in the old X-rays, showed up as definitely simian when looked at with newer machines.

The associated animal remains from Piltdown, the many teeth and the few bones, as also the various flint "implements," were closely studied, and all were readily eliminated as true artifacts. The most important of the flints—the four supposed paleolithic "tools"—bore a similar coating, believed to be natural stains from the soil, an unusual patina of yellowish brown. As it proved, this coating was easily washed off with hydrochloric acid. In addition, one of these "worked flints" yielded appreciable traces of chromium.

Among the mammalian remains, it was determined that two of the specimens had been derived from foreign sources. The elephant molar, judging by its uranium content, had almost certainly come from an anthropological dig in Tunisia—specifically a site near the town of Ichkeul. Examples of fossil teeth from Ichkeul all exhibited the same high radioactivity count as the Piltdown molar, unmatched elsewhere. More technically, in the elephant tooth, between the enamel and the cementum there was detected exactly the same ratio of radioactivity.

Also certainly derived from outside England was the hippopotamus tooth. Its combination of high ash content with low counts of fluorine and organic matter strongly indicated that it had been dug out of a limestone cave deposit, and this suggested the calcareous caves of the Mediterranean, especially on Malta. Other hippo teeth from the Malta caves carried much the same internal signature as the tooth from Piltdown.

The last of the specimens, the strange bone implement so like a cricket bat, which had been called unique in paleontological discovery and the most remarkable find made at Piltdown, quickly fell before the peering eyes in the laboratory. The slab was indeed a real fossil chipped from a primitive elephant's thigh bone. But it could never have been whittled while fresh, as experiments proved, and certainly not with a flint knife. Only on bone in a fossil state, where the texture is almost that of chalk, could the cutting have been accomplished at all, and then only by using an even-edged metal blade. No ancient flint knife could have managed the fine cuts and slices that had shaped the slab's narrow and pointed ends. In addition, after the ends had been sculpted by a keen-bladed modern knife, the usual iron staining had been deftly applied, giving it the necessary long-buried appearance.

By May 1954 the investigation, remarkably thorough and detailed, was complete. The results were written up in another special museum bulletin, this one running sixty oversize pages and heavily illustrated: *Further Contributions to the Solution of the Piltdown Problem.* Copies were ready for distribution by the end of June, and on the thirtieth of that month at a regular meeting of the Geological Society of London, in the same room where Piltdown Man had begun his official existence, the remarkable fraud was at last brought to an end. No one mourned the creature's departure. The elimination of Piltdown, all happily concurred, greatly simplified and even strengthened the general theory of human evolution.

A telling and provocative comment on the dramatic conclusion of the affair was soon provided by an American anthropologist, S. L. Washburn of the University of Chicago, which served to ignite a secondary disagreement as to the fundamentals of the scientific outlook. It was a brief dispute but with peculiar relevance for the Piltdown debacle. The solution of the English enigma, Washburn felt, marked nothing less than "the end of an era" in the study of man's origins:

> Present thinking about human evolution follows channels which were largely determined when there were few fossils, and few techniques. Now there are enough fossils so that any major theory of human evolution should not be based on specimens which do not include at least part of the brain case, some face, and jaw. This modest requirement would have prevented the vast waste of effort which resulted from the Piltdown fake. The great lesson of Piltdown for the student of human evolution is that *there never was enough of the fossil to justify the theories built around it* . . .
>
> The elimination of Piltdown . . . clarifies the origin of Homo sapiens, and emphasizes the importance of basing theories on adequately preserved and dated fossils. Above all, it stresses the importance of studying the originals with all the techniques available. The nitrogen analysis alone is more conclusive than all the opinions of the experts who studied the Piltdown fake.

On their face, Washburn's comments carried the conviction of common sense. Yet he was promptly challenged, and on quite reasonable grounds, by scientists insisting that partial and fragmentary fossil specimens must *not* be ignored or made to wait interminably for supporting evidence. E. A. Hooton of Harvard University perhaps expressed it best:

> I do not agree that anthropologists should refrain from formulating theories of human evolution around incomplete and fragmentary fossils. If Dubois had not been willing to speculate about the meaning of the calva and the supposedly associated femur of *Pithecanthropus*, an

important phase of human evolution that was accurately forecast from this discovery might have had to await the discoveries of Von Koenigswald nearly a half-century later . . .

Anthropologists need not be rash and irresponsible in the interpretation of fragmentary evidence, but they should not be pusillanimous and motivated principally by caution and fear of being proved wrong . . . persons who in science or in any field of thought or activity stand in perpetual fear of being "wrong" are never really right . . . The great lesson of the Piltdown business for me is that it is unwise to accept current scientific decisions and "proofs" as final, irrevocable, and conclusive, no matter how authoritative they may sound or look.

The disagreement, of course, frames a debate on matters of procedure which still today has not, and perhaps never can be, settled.* Both men were right. It becomes a finely graded matter of knowing when to do what, always making allowance for quirky human nature. In the case of Piltdown, as will be seen, those quirks played the pivotal role.

While the second phase of testing was under way, through the spring of 1954, Weiner and Oakley had the stimulating experience of corresponding with, then interviewing one of the original Piltdown excavators, Fr. Pierre Teilhard de Chardin. Now seventy-four and a paleontologist of world standing, he was based in New York City, associated professionally with the Wenner-Gren Foundation. His first personal comment on the news of fraud at Piltdown came in a reply he wrote to an inquiry from Oakley in which, curiously, he betrays no sign of shock. "I congratulate you most sincerely on your solution of the Piltdown problem," he declared. "Anatomically speaking, '*Eoanthropus*' was a kind of monster . . . Therefore, I am fundamentally pleased by your conclusions."

But unexpectedly, Teilhard also confessed that he found the whole startling idea of a deliberate hoax to be downright improbable—perhaps in reality it had all been the result of some strange accident. "Don't forget three things," he suggested briskly. First, "the pit at Piltdown was a perfect dumping place for the neighboring farm and cot-

* No reasonable thinker will deny that evolutionary theory today, in its details and overall, still tends to outrun the available evidence. A good grasp of that fact will help make clear how Piltdown could have succeeded so well for so long.

tage." Then also, "during the winter the pit was flooded." Third, "the water, in Wealdian clays, can stain (with iron) at remarkable speed." Under such conditions, he asked, "would it have been impossible for some collector who had in his possession some ape bones, to have thrown his discarded specimens into the pit?" Given all the ingenious secret labor that had obviously been expended in altering the bones, that was an extremely naive suggestion for Teilhard to make. In his reply, Oakley gently pointed out the utter impossibility of such a casual source for doctored bones.

(Oakley might have added what has never yet been well understood, and which deserves mention here. Because of its location, the pit at Piltdown could never have served as any sort of rubbish dump. It lay in full view close beside the estate's main road leading to the large manor house, which stood only some thirty or so yards from the pit in a straight line. It is surprising, in fact, that digging of any sort, even for road-mending flints, was permitted in so central and conspicuous a spot on the estate. The raw, open pit actually made an unsightly scar along the road's grassy border.)

In August of 1954, Teilhard left New York for a European trip, stopping for several days in London and finding time in his busy schedule for a two-hour visit to the Natural History Museum. Oakley had eagerly anticipated the meeting, since it gave him a chance to gather from a rare primary source firsthand information about Piltdown and the people directly concerned. But his hopes were rudely disappointed. Sitting in Oakley's office, the diffident Teilhard wished only to talk of current paleontological matters, in Africa and the rest of the world. Whenever the topic of Piltdown was broached he ventured only the briefest comment. "Sore subject!" he exclaimed at one point as he brushed the subject aside.

At least once during his stop at the museum Teilhard repeated his naive notion that the Barkham Manor pit had served as a "rubbish dump," and that the fossils might have been innocently thrown down by some passerby. In an apparent effort to excuse and account for his taciturn manner on Piltdown, he explained that at the time in question he had been "a mere youth . . . little more than a boy." But Oakley and his colleagues, having previously collected biographical data on their guest, knew very well that in 1912 Teilhard had been no boy, but an ordained priest of thirty-one.

Teilhard's visit to the museum proved so unsatisfactory to the English scientists that one of them, Le Gros Clark, was led to observations which neatly summed up the general impression that the priest left behind. "I am frankly puzzled by his attitude. If T. de C. had no hand in the forgery, why should he be *embarrassed*? If he knows something about the perpetrator (but was not himself implicated), surely he could say so—or at any rate make clear that he is not able to divulge matters which he regards as confidential? . . . I cannot imagine him ultimately planning the forgery."

A fitful correspondence continued through the year between Oakley and the priest, leaving the Englishman, along with his colleagues, vaguely disturbed as to Teilhard's true relation to the Piltdown events. Then in April 1955, in a friend's apartment in New York City, Teilhard died suddenly of a heart attack. He was the last of the original Piltdown figures to succumb: three months before, on January 7, Sir Arthur Keith had also died. Thereafter several decades would pass while each man's scientific standing—and in Teilhard's case, philosophical—continued to mount. In the end, however, and despite what appeared to be secure and lofty reputations, neither man would manage to escape the unrelenting glare of the Piltdown spotlight in its search for the guilty party.

Left unresolved in 1954 was the question of the Piltdown skull's true age, both jaw and cranium. The final Weiner-Oakley report had been content to designate the age loosely as "recent," or the equivalent, "post-Pleistocene," at most a few thousand years. No more precise results were obtainable then from direct chemical testing. Radiocarbon dating (carbon 14) was in use, but the process consumed much more material than could sensibly be spared from the bones. Refinement of the technique over the ensuing five years, however, proceeded so rapidly, requiring so much less material, that it was decided to use it on Piltdown. In the summer of 1959 the tantalizing question of Piltdown's actual age was at last settled. To many observers the official report, published in July in the journal *Nature*, came as something of a surprise.

The cranium yielded an age reading of 620 years, with a plus/minus factor of one hundred years. The skull's owner, in other words, had died probably in the fourteenth century A.D., but no earlier than the thirteenth. The jaw proved to be a bit younger, only five hundred years, placing it in the fifteenth century. Yet because of the plus/minus factor of one hundred years, it was also possible that jaw and cranium were about the same age, or even that the jaw was slightly the older of the two pieces. The age difference, in any case, provided no new information.

In addition, the jaw was now definitely identified as that of an orang-utan, thus ultimately deriving from Borneo or Sumatra. Many possible ways existed in which the forger might have obtained it, said the report, including, most obviously, by purchase from "a dealer in ethnographical materials." The Piltdown researchers themselves quickly pointed to an old and well-known collection of orangutan bones brought home to London in 1875 by the zoologist A. H. Everett. A part of that large and varied collection had been placed in the Natural History Museum, and in the time of Piltdown it was still there in storage (a check of the original inventory by Oakley showed that none of the skulls was missing). But a sizable portion of the collection had been dispersed without record and had simply disappeared: "What became of it is unknown. Did it pass into the hands of dealers?"

Such orang skulls need not have been very difficult to locate, added the report, even apart from dealers, or ground burials, the usual source. Villagers in Borneo, and probably in Sumatra, had long made a habit of keeping whole orang skulls as fetishes or trophies, valuable properties that were religiously passed down in families. Some orang skulls, it was known, had hung in village longhouses for twenty generations, perhaps four hundred years or more.

The provenance of the human skullcap was more easily explained. It could have come from almost any very old cemetery in Britain, either directly or, again, through the willing hands of some dealer in fossils (in this instance, perhaps, a none-too-scrupulous dealer, since he would have been handling more contemporary remains). "Presumably they were fragments of a skull," suggested the report, "selected on account of its unusual thickness from among a series obtained in the excavation of some ancient burial ground." In England such ancient grounds were plentiful.

Piltdown Man, the most famous creature ever to grace the prehistoric scene, had been ingeniously manufactured from a medieval Englishman and a Far Eastern ape.

6

Pursuing Dawson

The High Street in Lewes, starting at the river end of town, runs uphill until it reaches the scanty but imposing remains of the old Norman castle at the town's center, where it begins to level off. The castle gate, lofty and grand, still occupies its original site, set a little back from the High Street and fronting an open space between rows of old buildings. Beside and behind the gate, where the castle itself once rose, can be seen only a small portion of the outer walls, but two of the original towers still stand, overlooking the picturesque sprawl of the ancient town. Running through the gate and across the broad, open summit is a narrow public street that passes by several private homes and a bowling green. Just inside the gate, to the right, is Castle Gate house. Opposite, standing impressively behind low walls, is Castle Lodge, once the residence of Charles Dawson and his family.

On August 8, 1953, Joseph Weiner walked up the High Street and turned in at the castle, where for the first time he viewed the looming gate, with Castle Lodge conspicuously in sight to its left. But Weiner had not come to Lewes to inspect the ruins or to see Dawson's old home. He was in search of the headquarters of the Sussex Archaeological Society, and had been told that it was located in Barbican House, hard by the old castle. There he found it, standing a few yards short of the gate, a squat, three-story building of red brick.

Encouraged by Le Gros Clark, his superior at Oxford, Weiner had not waited to begin his search for the unknown Piltdown forger. Only a week had passed since the evening of the London banquet, and it was scarcely a day or two since he and Clark phoned Oakley with their suspicions. Inevitably, but quite reasonably, all three had concluded that Dawson himself, the instigator and driving force at Piltdown, offered the prime target for investigation—how little anyone really knew, it was agreed, about the Uckfield solicitor! Within a matter of only ten days or so, after Weiner had twice visited the Lewes area, they would know a great deal more.

The first indication of something awry came almost as soon as Weiner entered Barbican House. Passing through the diminutive doorway, he saw that the ground floor held a tidy little museum, its many exhibits mounted with professional competence and flair. Here, as he said later, he naturally expected to "see much made of Piltdown." On the contrary, he found that the various display cases were bare of anything relating to Dawson's discovery, beyond a single cast of the skull. Even this cast, Weiner saw on reading the card, had been a late and unofficial addition to the museum, a private donation by a member in 1928. "Nothing else of the Piltdown assemblage was on view," he noted in considerable surprise.

On the third floor in several small, ramshackle rooms was located the society's well-stocked research library. Scarcely a half-hour's study in its crowded reading area provided Weiner with even more reason for puzzlement. In the multivolume *Sussex Archaeological Collections,* an annual publication describing everything of antiquarian interest occurring during the year in southern England, he searched for references to Dawson, again feeling sure of extensive coverage. But nowhere in the six thick volumes for 1911–16 was there the slightest mention of Piltdown: no description or discussion of the great discovery, no record of any meeting on the topic or of Dawson being invited to address the society, a regular custom. No recognition of Dawson's newfound fame could be found, no reference to his death. Except for a single instance, in the years *after* his death, Piltdown went unmentioned in the *Collections.* Only in 1925 had there been a talk given before the society on Piltdown, delivered by Arthur Smith Woodward. But the text of the talk, despite the identity of the speaker, had not been printed in the *Collections,* an unusual omission.

For almost two decades, beginning in 1892, Dawson had been active in the society, eventually becoming by far its most prominent member. Yet in the days of his fame, with his name and deeds familiar among world scientists, and through all the years afterward, his own society had ignored him. Later, Weiner confessed himself "astonished" by what he decided could only be "deliberate avoidance" of Dawson and his work among his local associates. It was a situation which cried out for explanation, and that same day at the home of Louis Salzman, a leading member of the society who had known Dawson well, Weiner's questions were answered. No word of the arresting truth had ever reached London, it seems, but for more than a dozen years before his death in 1916 Dawson had been virtually ostracized in Lewes, at least in its antiquarian circles. It had all begun, explained Salzman, because of the underhanded manner in which, just before his marriage, Dawson had acquired ownership of Castle Lodge, which for long had been the Society's headquarters.

In 1903, Dawson, unmarried, lived in Uckfield, occupying rented rooms a few blocks from his law offices. At that same time the Sussex Archaeological Society was happily in possession of Castle Lodge, where it had been located for twenty years, and where it fully expected to remain. Then, that fall, the society's governing council received an official notification that the lodge had been sold. The society, museum and all, must vacate the premises. The new owner was Charles Dawson, who had purchased the building for his own use.

The society's utter "consternation" at the unexpected news was all the greater since it had been well understood that, if the lodge was to be sold, the society would have clear rights of first refusal. But no one in the organization had known anything of an impending sale, and it was soon found that Dawson's management of the purchase had involved outright misrepresentation. Until the last moment, the seller had assumed that Dawson was acting on behalf of the society, a mistake fostered by Dawson's being a lawyer and by his use of notepaper bearing the society's letterhead. That the damning story rested on more than gossip or hearsay, Salzman readily demonstrated by showing Weiner a notice that had appeared in the *Collections* of 1904:

The Council, in the autumn, received an intimation that the Castle Lodge, which had been occupied by the Society since the year 1885, had been sold to Mr. Dawson, and a notice to quit at midsummer 1904 had been served by him on the Secretary. This purchase by one of our own members, and its consequences, took the Council completely by surprise—as it understood that if the property was to be sold the Society should have the option of acquiring it.

Plans were being made, added the notice, for removal to temporary quarters. But for members and for the society's ongoing business there would be "a certain amount of trouble and inconvenience." Eventually, temporary quarters were found in a building a few blocks farther down the High Street.

In January 1905, Dawson, aged forty, married Helen Postlethwaite, a wealthy widow of London's affluent Mayfair section. There occurred a delay in the couple's occupying Castle Lodge, in part because of extensive remodeling, but by early 1907 they had moved in. Soon after, the archeological society, wishing to renew its physical association with the ancient castle, rented quarters in Barbican House (where it remains today), distant from Castle Lodge a matter of yards, and in full view of its windows. From then on, as Weiner noted, life for the Dawsons meant encountering "the daily coolness of the recently evicted tenants . . . altogether a socially trying environment."

From his own experience, the elderly Salzman—he was then seventy-five and would live into his nineties—supplied another distasteful instance of what he claimed was Dawson's chicanery. In 1910 Dawson's large, two-volume work, the *History of Hastings Castle*, was published by Constable in London. Salzman reviewed it briefly in the *Collections*, expressing a decidedly harsh opinion. Dawson, he admitted, had shown great industry in gathering the valuable material "but little judgment in its selection and arrangement." Even more damningly, he went on, there were many errors of translation and numerous misreadings: "It is difficult to say how far these are due to carelessness, inaccuracy, and neglect of proofreading, and how far to reliance upon secondhand authorities, as references are frequently omitted or given in an unintelligible form." He then added a charge that for Dawson would have carried a particular sting: "In many cases when matter is taken, mistakes and all, from earlier writers no acknowledgement of the source is made."

In retaliation for this starkly unfavorable notice, said Salzman, Dawson had gone so far as to try derailing his application, then pending, for admission to the Society of Antiquaries in London. Candidates for the coveted honor were not permitted to lobby or to indulge in any sort of campaigning, and the least effort to influence the selection committee was cause for rejection. The angry Dawson, knowing this, had deliberately sent anonymous postcards to key members urging Salzman's election. If the trick had not been uncovered in time, explained Salzman, and disavowed by him in a letter, his application would have failed. That Dawson was behind the ugly tactic, Salzman couldn't prove, but insisted he had good reason to believe.

Concluding his interview with Weiner, Salzman suggested that his visitor pay a call on another Lewes resident, the county's assistant surveyor, A. P. Pollard. A good deal more of interest about Dawson and Piltdown, Salzman hinted, could be gleaned from that quarter. Weiner found Pollard in his office and explained that he was interested in hearing about Piltdown, but said nothing about fraud. Abruptly, as Weiner recalled, Pollard inquired whether he had "any reason to distrust the discovery?" Intrigued by the implication of the question, Weiner admitted that, in general, he had. "I am not surprised," responded Pollard. "I believe it to be a fraud. At least that is what my old friend Harry Morris used to say."

Another resident of Lewes, Harry Morris was a bank clerk and an ardent amateur paleontologist, as well as an acquaintance of Dawson's. He had died in 1940, but not, it developed, before he had several times privately accused Dawson of faking the Piltdown flints, and much else left vague. He had even written down his accusation, added Pollard, notes he had seen with his own eyes. They were in an old flint cabinet left to him by Morris, which was where he first came across them several years before. But he no longer had the cabinet. Not caring about flints, he had traded it to a collector in Ditchling, a town just west of Lewes. The man's name was Wood. He was dead but his widow might still be alive. About that Pollard couldn't say.

The week that followed Weiner's first trip to Lewes was taken up by his extensive duties at Oxford. But he used the time to ascertain that a Mrs. Fred Wood was indeed still living in the village of Ditchling, still had her husband's old cabinet, and would be happy to show it. Early the next Saturday, August 15, Weiner with a companion stood in Mrs. Wood's parlor staring in fascination at a plain, wooden, four-foot-high chest of drawers.

He began his inspection with the top drawer—there were twelve in all, he counted—carefully pulling out each one in turn, "our anxiety increasing as we proceeded." Then as the bottom drawer came out he spotted what appeared to be two small cards, one of which was lying beneath a chunky, four-inch flint. Not only on the cards but also on the flint itself there was writing.

The card beneath the flint read: "Stained with permanganate of potash and exchanged by D. for my most valued specimen!—H.M." Directly on the flint, over two of its flattened planes, was scrawled in fading ink, "Stained by C. Dawson with intent to defraud (all).—H.M." The reverse of the card held a quotation identified as from *Macbeth*. It is headed with the bare words "Dawson's Farce":

> Let not light see my black and deep desires,
> The eye wink at the hand; yet let that be
> Which the Eye fears when it is done—to see!

The second card showed the lengthiest message and the most challenging inscription: "Judging from an overheard conversation there is every reason to suppose that the 'canine tooth' found *at P. Down was imported from France.* I challenge the S.K. museum authorities to test the implements of the same patina as this stone which the imposter Dawson says were 'excavated from the Pit!' They will be found [to] be *white* if hydrochlorate acid be applied. H.M. Truth will out." The note, in ink, is written across some words scrawled in pencil: "Watch C. Dawson. Kind regards."

That was all. Nothing in the cabinet helped settle the dates of the various notes, or even if all had been written at the same time. If it were not for the phrase penciled beneath the longest message, "*Watch C. Dawson,*" which seemed to speak of the living man, the notes might be assigned to any year up to Morris' death in 1940. More puzzling yet was the question of purpose. Why had Morris written them at all if they were to remain hidden away in a cabinet? Whom did he have in mind for those "Kind regards"? Most tantalizing of all was the unexpected mention of the canine as coming from France—a link to Teilhard? wondered Weiner—and its cryptic aside concerning an "overheard" conversation. The Morris intimations, it appeared, if they were to be taken seriously, spread higher and wider than mere doctored flint implements.

For all these questions, at that moment Weiner could imagine no adequate answers. With the permission of Mrs. Wood, he pocketed the two cards and the written-on flint, borrowing them for further study (later they were purchased by the museum).

From Ditchling, Weiner drove the short distance back to Lewes, where he paid a curious visit to Castle Lodge. Cordially welcomed by the tenant, Dr. Nicholl, he soon heard something that seemed nearly as arresting as the contents of the Morris cabinet, at least at first. In a letter to Oakley, he explained that it involved the accidental finding years before of some extraneous skulls. "We had one episode of the most fantastic and exciting kind, which fizzled out," he wrote. ". . . It was in discussion with Nicholl jnr. that he suddenly said that he remembered as a boy of 8 that while the floor in the back dining room was being relaid for dry rot 'two monkey skulls were found under the floor.' " Disappointingly, however, on the following day "no one could be found to corroborate" the story. Weiner spoke with Nicholl, Sr., with the family's old nurse, and with the gardener, even found the foreman who had supervised the repair work. None could remember any talk of monkey skulls. Undiscouraged, Weiner reported that "Nicholl jnr. is keen to help us, and will allow us to search the large lofts in the house at any time."

Neither at the initial announcement of the forgery in November 1953, nor at the full revelation that came the following June, was there any mention by the museum of a possible suspect. The question of the perpetrator's identity was simply not addressed. But as the *Times* had quickly commented, "Who did it? is a question many will ask," and with that the paper proceeded to focus the spotlight squarely on Dawson. It was a decision, natural enough in the circumstances, repeated on the front pages of paper after paper.

Sketching the story of the original finds, the *Times* showed how Dawson originated the search, being joined only later by Woodward and Teilhard. Here, said the paper, in the persons of the scientist and the priest, were two witnesses "of the highest character" who had aided Dawson, and it was hard to avoid the idea that the jaw and the canine had been planted by some third person in order to have the forgeries "unimpeachably discovered." With that, the paper came very close to leveling a specific charge: "If that third person were to prove to be

Charles Dawson, it would be but one more instance of desire for fame (since money was certainly not here the object) leading a scholar into dishonesty."

Immediately, from Dawson's family and close relatives there came a strenuous protest, even hinting at legal action. His stepson, Captain F. L. Postlethwaite, sent a letter to the *Times* that was published only two days after the initial report of the forgery. While he admitted having been away on military duty in the Sudan for much of the Piltdown period, he rejected with "considerable indignation" the implied charge against his stepfather:

Charles Dawson was an unassuming and thoroughly honest man . . . Until the discovery at Piltdown he did not display any particular interest in skulls, human or otherwise, and so far as I know had none in his possession. To suggest that he had the knowledge and the skill to break an ape's jawbone in exactly the right place, to pare the teeth to ensure a perfect fit to the upper skull, and to disguise the whole in such a manner as to deceive his partner, a scientist of international repute, would surely be absurd, and personally I am doubtful whether he ever had the opportunity of doing so.

No—Charles Dawson was at all times far too honest and faithful to his research to have been accessory to any faking whatsoever. He was himself duped . . .

That same day in London's *Daily Herald* appeared an interview with Dawson's nephew and two nieces, children of his brother Trevor. Their admired uncle, they declared, "was not the type of man to hoax anybody . . . we were proud of our uncle. If anyone suggests he was a party to the hoax we shall certainly do something about it to clear his name." However, while the newspapers continued to spotlight Dawson as the likely culprit, his family failed to rally, and they were not heard from again in public.

Inevitably, the newspaper publicity brought Weiner any number of individuals claiming to possess privileged information, and some of this appeared at first to have value. There was, for instance, Robert Essex, a retired schoolteacher of Uckfield and an old acquaintance of Dawson's. He promised to deliver inside information but, on a visit to the museum, he spent the hour insisting at great length on the guilt of Teilhard de Chardin. His story proved to be a rambling patchwork of suppositions, and in addition "he was nearly stone deaf and his hearing apparatus was out of order . . . he was pretty hopeless on dates."

Another old Lewes resident and friend of Dawson's, Earnest Clarke, momentarily captured Weiner's attention when he told of once seeing bones in Dawson's cellar. While a dinner guest with his wife at Castle Lodge, well before the approach to Woodward, Dawson had invited Clarke down to his cellar. Supposedly—and for reasons neither stated nor obvious—he showed him "several pieces of bone. They were dry." Closely questioned by Weiner, Clarke could not recall if the cellar had the look of a workroom, could not remember seeing "bottles of chemicals or tools." Eventually, Weiner learned that Dawson, before going to Woodward, had shown some of the Piltdown cranial fragments to several people in quite a casual manner, including Teilhard.

Much more arresting for both Weiner and Oakley was some information about Dawson that reached them from a Captain Guy St. Barbe, also a former resident of Lewes, who was then living at St. Albans, to the north of London. After several interviews with St. Barbe, at the museum and at his home in St. Albans, the two felt they were at last in possession of some veritable evidence concerning Dawson's role as the forger. Like Clarke, the captain claimed to have actually seen something in Dawson's hands, but potentially far more incriminating.

St. Barbe, then in his seventies, during the Piltdown excitement had lived near Lewes (the title "captain" was never explained but it was assumed that he was a retired military man). He had known Dawson both socially and as one of his law clients, he said, and on two particular occasions had gone to consult him on business. Both times he had entered Dawson's Uckfield office unannounced and without knocking, in his exuberant way had "burst in." Each time he had surprised the solicitor in the midst of "staining" things, the first time bones, the second flints. Spread on a table, he recalled, were many small "dishes," each holding a "brownish liquid," while in the air lingered a "strong smell as of a chemist shop," the odor of iodine being pronounced. At the unexpected intrusion, said Barbe, Dawson became "very agitated" and was unable to hide his "extreme embarrassment." Lamely, he had explained the curious array of dishes by saying he was studying how discoloration of specimens took place in nature.

Suspicious, yet uncertain of the full significance of what he saw, St. Barbe had said nothing to anyone. Shortly afterward at a garden party, he explained, he met another Lewes man, Major Reginald Marriott, a much-decorated officer of the Royal Marines, then retired. Both confessed to being avid followers of Sussex archeology, and soon they were talking of Dawson and Piltdown, upon which St. Barbe decided to tell of the staining incident. Instead of voicing shock or even surprise, Marriott shook his head. He, too, had once entered Dawson's office unexpectedly, he confided, and "had also seen the staining in progress." Exhaustively, the two discussed the solicitor's odd behavior, finally concluding that for the time being they would maintain silence. They would wait until doubts had been raised by qualified observers, the scientists themselves, then would volunteer what they knew. But somehow, as the years passed it had all just seemed to fade away. St. Barbe, removed from the scene by his service in World War One, never returned to Sussex. In 1930 Marriott died.

As Weiner realized, the timing of the three intrusions into Dawson's office was crucial. All of the visits, it seemed, had taken place in the summer of 1913, *after* the initial public disclosure of the discovery, but *before* the finding of the other artifacts (possibly the canine, certainly some of the flints and animal remains, and the bone implement). Sufficient evidence for fixing the date was St. Barbe's own clear memory of having seen in Dawson's office a cast of the skull, not distributed until May 1913. By fall 1914, with the start of the war, St. Barbe was gone from Sussex.

Further convolutions turned up around the St. Barbe testimony when Weiner located Marriott's daughter, who was still living in the Lewes area. Readily the woman explained that "it was well known in his family" how her father had for long viewed Piltdown as an outright fraud. Next, it developed that Marriott had been a close friend of Harry Morris, and the two had often rehearsed all they knew that was questionable about Dawson and Piltdown. During the years of Piltdown's ascendancy, it began to seem, at least three men of Lewes, all of them acquainted with Dawson, had steadily held back the evidence that might, if revealed, have begun the perpetrator's downfall. "Their hesitancy is understandable enough," reasoned Weiner. "One must realize the enormous weight of authority which buttressed Dawson and his discovery . . . famous biologists . . . foremost anatomists . . . leading geologists . . . all had given their massive verdict in Woodward and Dawson's favor." Not surprising, in the face of such impressive support, was the reluctance of the amateurs Marriott, Morris, and St. Barbe to make an "appalling allegation" against Dawson.

(Strangely, it did not, then or afterward, strike Weiner or any of his colleagues to think how inconsistent, even self-contradictory, was their ready acceptance of all that St. Barbe told them. Here in the person of Charles Dawson, supposedly, was a forger admitted by all to be highly ingenious and daring, a cunning planner possessed of masterful skills in the execution of his dark designs. Yet here he is seen ignoring the most ordinary precautions, being content to use his own very public law office as his secret laboratory—and then had not sense enough to lock the door when working! So much should, by way of anticipation, be mentioned here.)

Through the spring of 1954 Weiner kept up his pursuit of Dawson, which now included a close study of all the solicitor's correspondence preserved by Woodward at the Natural History Museum. Some ninety letters and notes were turned up in the old files, written by Dawson to Woodward from 1909 to his death in 1916, all concerned with paleontological topics. Disappointingly, it proved to be a one-sided correspondence, for Woodward had kept no copies of his own letters to Dawson (his customary practice, apparently). Still, what was available allowed Weiner to sketch out, for the first time in some detail, the true progress of the work and discoveries at Piltdown. Culling other sources—libraries, museums at Hastings and Brighton, files of journals and newspapers, various informants in Lewes and elsewhere—he was able to form an intriguing picture of Dawson's remarkable thirty-five-year career as an amateur archeologist-historian.

In addition to making himself an expert on Wealden dinosaurs and mammals, the busy Uckfield solicitor had also emerged as an authority on aspects of ancient Sussex life and industry, including ironwork, pottery, and glassware. On all three subjects, each a major discipline in itself, he wrote learnedly and at length, as well as collecting related artifacts and supervising major exhibitions. His personal collection of antiquities, painstakingly gathered over many years at no little expense and long exhibited on loan to the Hastings Museum, included rare items of jade and bronze, as well as exceptional pieces made of stone, iron, and glass. At various periods he had helped in the excavation of important Sussex discoveries—one instance involving some Neolithic skeletons, another some rare caves full of a mixture of Roman and medieval artifacts. Antique furniture captured his interest equally with old etchings, paintings, and maps. He was associated with the discovery of a pocket of natural gas at Heathfield, which was used for a long time to light the town's railway station. He reported an incipient horn growing on the head of a common cart horse, and he spotted an extremely rare thirteenth vertebra in a human skeleton, which he felt identified a new race. He even attested to the presence of a sea serpent in the English Channel, spotted by himself when crossing on the ferry from Newhaven to Dieppe.

It was while listing all these disparate items and others that Weiner became aware of a common thread winding through much of the extraordinary assortment: "It is not merely the novel or spectacular element which is characteristic, so much as the persistence of his concern for the 'transitional,' or to use his favourite term, the 'intermediate' form. Such in fact were the things which brought him the notice of palaeontologists and archaeologists alike." In the years before Piltdown, noted Weiner, Dawson had been associated with such "transitional" artifacts as a boat that was half coracle and half canoe, a horseshoe that was both tied and nailed on, a Neolithic stone weapon with an unprecedented wooden haft, the first use of cast iron in Britain, a mammalian form between *Ptychodus* and *Hybocladus*, and an unusual Norman prick-spur. That he should readily have appreciated the full meaning and implication of a Dawn Man "emerges irresistibly from the record of his activities, his abilities, and his habits of thought."

Dawson's historical writings, varied and impressive, contributed to a number of academic journals, were closely inspected, as was his massive, two-volume *History of Hastings Castle*. Many of these studies, it soon became clear, were hardly more than "useful compilations," taken from older printed sources, often without acknowledgment. Large portions of the *Hastings Castle*, as Weiner eventually found, had already been traced by competent authority to a particular source, the unpublished, century-old manuscript of another Sussex archeologist. Dawson's acknowledgment of that debt in his book's preface, agreed Weiner, was at best feeble and at worst misleading. The book, which by then had become a standard reference work, was judged to be "less a product of genuine scholarship" than the clever result "of extensive plagiarism," a conclusion that reached Weiner from a curator at the Hastings Museum who had compared Dawson's text with its unpublished source. But here the trail became blurred, for some few other observers, more generous, defended Dawson in the matter of his Hastings Castle volume. The citation of the earlier author in the book's preface they saw as being quite full and frank, fairly justifying whatever use Dawson cared to make of it.

Very early in his Piltdown involvement, Weiner had determined to write a book on the outcome. The manuscript, rather hastily prepared in the midst of his ongoing investigation and his academic duties, was virtually complete by the summer of 1954. Published six months later by Oxford University Press, it was a small volume called simply *The Piltdown Forgery*. Supplying an interested public with the first extended and detailed treatment of the affair, starting with the original discoveries in 1912 and before, it was well received by the scientific community. But concerning the identity of the perpetrator the book cautiously held back, its conclusion neatly balanced on the fine edge of probability (a little influenced, perhaps, by those implied threats of legal action from Dawson's relatives).

Unable to counter or reject the claim that it was the workmen digging at the pit who first uncovered the skull, promptly smashing it, Weiner accepted what appeared to be the obvious solution. The perpetrator, he thought, had simply planted the whole intact cranium, or brain case, of *Eoanthropus*—the so-called "cocoa-nut"—for the diggers to find. Concerning what happened next he wasn't quite sure. Perhaps the pieces of the broken cranium had been handed by the diggers to the estate's tenant, Mr. Kenward (his daughter claimed as much), who passed them to Dawson. Or the few early pieces might have reached the solicitor's hands directly from the workmen. In each case, he thought, the greater part of the cranium would have remained in the ground for later discovery.

Weiner also toyed with the idea of a substitution of skulls: an innocuous specimen was planted for the diggers to find, with the real *Eoanthropus* fragments later taking its place. That attractive solution, however, had problems of its own. For example, the diggers might have had sense enough *not* to smash it, but to deliver it intact.

As to Dawson's outright guilt or innocence, nothing further had come of the cryptically accusing Morris notes, or the uncertain testimony of St. Barbe. As a result, Weiner at the last felt unable to make a definite pronouncement, and had to content himself with strong implication. If the solicitor had been nothing but a dupe, he stated carefully, it was remarkable how perfectly his actions coincided with those of the perpetrator. All that was known, for instance, about the delicate operation of the staining of the specimens was a direct indictment of Dawson: no explanation of the timing and sequence of the staining was enough to "clearly dissociate Dawson's actions from the perpetrator's." If he had been himself a victim of some "unknown manipulator," why were his own reports of the earliest stages of the discovery so vague and confusing? Why had he told Woodward so little about Sheffield Park and Barcombe Mills? If he were not himself directly implicated, then who was the Machiavellian operator who had kept him under such close watch for so many years, who could predict and, it appeared, often direct his movements?

Judiciously, Weiner summed up: "We have seen how strangely difficult it is to dissociate Charles Dawson from the suspicious episodes of the Piltdown history. We have tried to provide exculpatory interpretations of his entanglement in these events. What emerges, however, is that it is not possible to maintain that Dawson could not have been the actual perpetrator." The rather convoluted expression was continued as he pointed out the difficulty "of accepting [Dawson's] known activities as compatible with a complete unawareness of the real state of affairs."

Sensibly unwilling to make an outright condemnation on the basis of the available evidence, all of it circumstantial, Weiner then unexpectedly went a step farther, sorely weakening his own case. "In the circumstances," he suggested, "can we withhold from Dawson the one alternative possibility, remote though it seems, but which we cannot altogether disprove: that he might, after all, have been implicated in a 'joke,' perhaps not even his own, which went too far? Would it not be fairer to one who cannot speak for himself to let it go at that?"

Unfortunately, in his commendable effort to be rigidly fair and objective, Weiner here backpedaled into an absurdity of his own. There is not the least possibility that Piltdown was begun or carried through as a merry-hearted prank.

There Weiner left it. For another twenty-six years he continued to live and work in London. But while he followed the developing story of Piltdown in private, once or twice lecturing on the topic, and though the tangle of facts about Dawson cried out for unraveling, he wrote no more about the fraud he had been the principal means of exposing.

7

List of Suspects

Only a little surprising is the way in which candidates for the role of the Piltdown forger have proliferated during the last several decades. Charles Dawson, though continuing as leading suspect, either as sole perpetrator or coconspirator, in the absence of solid evidence has managed to avoid a final verdict (indeed, he has by some been declared innocent, seen as no more than a dupe). As a result, accusations have been lodged against almost every name connected with the dramatic event, and nearly all imaginable forms of conspiracy have been suggested. One elaborate theory pictures a shifting alliance of no less than half a dozen unscrupulous plotters. Favored by many at the moment is an arrangement which has some shadowy figure masterminding events, usually with Dawson as henchman, willingly or by coercion.

Of the twenty or so names so far put forward, almost half are mere fleeting, unadorned guesses, bare of anything resembling evidence. These, devoid of interest as they are, may be dismissed without mention. Eleven other candidates with at least a modicum of justification, though the evidence against them is hardly more than circumstantial, have been written about at more or less length. Out of the eleven, eight rest in varying degrees on claims that must be accounted little better than tenuous. Only three of the eleven demand extended examination: Conan Doyle, Teilhard de Chardin, and Sir Arthur Keith, and each of these must be given a whole chapter to himself. Before taking up those three separately, the weaker eight may be allowed to pass by in brief review.

Two men from the Lewes-Hastings area stand accused, ordinary citizens, neither of whom was directly active at Piltdown. One was long employed as a museum curator at Hastings: William Butterfield (died 1935). The other was a jeweler by profession, far better known in his role of amateur paleontologist: Lewis Abbott (died 1933). Both men had been acquaintances of Dawson, if not quite friends.

Against Butterfield the case hinges on a revenge theme, the supposed ill-will and choleric anger he at one time felt toward Dawson. Entirely on his own, it is charged, he created the whole Piltdown assemblage, then manipulated Dawson into the position of discoverer, all as a way of avenging himself on his foe. The cause of the curator's anger is given as an incident, a real one to be sure, that occurred in the spring of 1909. In a Hastings quarry that summer, some dinosaur bones (Iguanodon) were uncovered by workmen, who informed Dawson. Quietly, without telling Butterfield, Dawson appropriated the bones, then sent them to the Natural History Museum in London, consigned to Woodward. Deeply hurt and angered, it is charged, resentful over losing his chance at the valuable specimens, Butterfield proceeded to take his revenge, using materials from his own museum at Hastings. The actual planting of the fakes in the pit at Barkham Manor, runs the theory, was accomplished not by Butterfield but by someone he recruited (how this was done is not addressed), an accomplice who had a good excuse for being on the spot, the workman Venus Hargreaves.

The charge made against Butterfield—who was known to have had a calm, even placid, disposition—is the classic instance of a serious accusation being built in its entirety on freewheeling speculation. The only pertinent facts contained in it are that Dawson did appropriate the dinosaur bones (quite legitimately), and that Butterfield did, once at least, express great annoyance over the loss to his museum. Beyond that, aside from an endless series of guesses, there is literally nothing to examine. In addition, just how much of crucial importance is overlooked and left out, how much is not even understood, will become obvious in the several discussions below.

More interesting than Butterfield, if equally as doubtful, is the figure of Lewis Abbott. The owner of a crowded jewelry shop in Hastings, he had known Dawson since the turn of the century, initially through the connection of both men with the Hastings Museum. Well respected by professionals for his wide knowledge of Wealden flora and fauna, especially its oldest gravels, Abbott was regarded "almost as an oracle" on the geology of southern England. From the start of the exposure he caught the attention of the questing Weiner, who gave him full coverage in his 1955 book but only hinting at his possible complicity as a co-conspirator of Dawson (perhaps as supplier of the skull). After Weiner, a few writers agreed with the possibility, some even the likelihood of his involvement, but only in 1986 did the old accusation become full-blown.

In *The Piltdown Inquest*, by Charles Blinderman, a full chapter attempts to demonstrate that the jeweler had "the best credentials to be the Piltdown hoaxer." But even thirty pages of background discussion, along with wholesale use of the possibility-probability approach, mixed with earnest but uncritical maneuvering among a myriad of circumstances, proved inadequate. No slightest actual sign, no smallest definite indication of Abbott's guilt resulted. Blinderman ends his study of Abbott by openly admitting to serious uncertainty over every important aspect of the case—materials, methods, and a motive. Yet he declares himself thoroughly convinced of Abbott's guilt.

In that contradiction may be discerned another peculiar feature of the hunt for the Piltdown perpetrator. Pure feeling and instinct are too often permitted to run rampant. People who ordinarily show themselves to be steady and level-headed, in treating Piltdown are for some reason apt to slip all restraint.

It is true that before the discoveries at Barkham Manor Abbott talked with Dawson, in general and more than once, about the occurrence of ancient, possibly fossil-bearing gravels in the Weald (as the two talked of many other antiquarian topics). It is also true that Dawson, especially in Piltdown's early stages, showed Abbott some of the bones and flints (as he showed the same things to any number of others). It is also true that Abbott later claimed credit for having "stimulated" Dawson's searches, insisting that without that stimulus "it is quite possible that these important things would never have been brought to light." But that is the claim of a man who believed in the *reality* of Piltdown, and in the end he remains what Weiner early labeled him, an egregiously self-important personality, "fiery, bombastic, inspiring and weird." He was convinced he had made paleontological discoveries of worldwide importance, but after his death his scientific reputation faltered badly, when much of his work was found to be defective, invalidated by later information or rejected as simply mistaken. In reality, Abbott was that saddest of figures, the intelligent, capable man with an unfortunate flaw, who never fully masters himself or his abilities. His was not the temperament from which master forgers are made, and he can be tied to Piltdown only if unsupported conjecture, without the least suspicious fact, is allowed a place.

Three leading scientists come next, two of them active and well known at the time of Piltdown, one a young unknown. Against none of the three is the case worth dwelling on. Against one, William Sollas of Oxford (died 1936), the charge assumes almost a bizarre, even a burlesque quality.

One of England's most eminent scientists in the century's early years, at no point had Sollas actually come within the Piltdown sphere. From time to time he wrote of it, as did many of his colleagues, and he was, to one degree or another, known to all the principals. But busy with his own pressing duties at the university and elsewhere, he had no hand in the excitement arising at Barkham Manor, direct or indirect, nor when the exposure came in 1953 did his name so much as receive mention. Only in 1978 did the accusation against him surface, more than forty years after his death at age eighty-seven. The charge was brought by a man who had been Sollas' assistant at Oxford, later his successor, Professor J. A. Douglas (himself without any connection to Piltdown). The ninety-three-year-old Douglas did not even make his charge in person, but through the medium of a tape recording, the tape being confided to a friend's care. After Douglas' death early in 1978 the existence of the tape was revealed, and it was played for a roomful of bemused scientists. In a lengthy, rambling discourse, the aged voice proposed Professor Sollas as the mastermind who planned and directed the work of Charles Dawson in erecting the hoax.

The supposed evidence supplied by Douglas consisted of nothing more than pure argumentative speculation. "I can remember as if it were yesterday," rumbled Douglas, how a mysterious packet of potassium chromate arrived at the school one day for Sollas. There was also a memory of Sollas "borrowing apes' teeth from the department of Human Anatomy," and there was the inevitable charge that Dawson must have had an accomplice, "someone who had higher scientific attainments." One short passage will give the flavor of the kind of loose assertion that issued from the loudspeaker that day:

> I know that Sollas knew Dawson and had visited him, but on how many occasions I cannot say for I was abroad at the time . . . When the skull was proved to be fraudulent it would appear obvious that Dawson had an accomplice. The bones did not originate there but were planted there and it is most unlikely that a Sussex solicitor would have had such bones in his possession. They must have been provided from an outside source . . . and it is possible that the thing started as a joke and then got out of hand . . .

Sollas and Arthur Woodward, Douglas went on, had been "bitter enemies," their feud triggered by Sollas' resentment of Woodward's pretensions to a leading role in British paleontology. Once, it seems, the irritated Sollas had even enticed Woodward into suffering scientific ridicule in the affair of the Sherborne horse, a bit of ancient bone bearing a thin sketch of a horse's head. Sollas knew that the tracing on the bone was spurious, said Douglas, and he purposely let Woodward make a fool of himself by officially describing it as authentic. Then Sollas sprung his trap (but ten years afterward, be it noted) by disclosing the truth about the Sherborne fossil, declaring that it was the work of two schoolboys. With that, Douglas' voice on the tape inquired whether it was not reasonable to believe that Sollas, having succeeded so well with one small fraud, was not "capable" of another, more ambitious hoax against the same man.

Some sincere and intelligent people were persuaded by the disembodied voice to consider, even to take seriously, what Professor Douglas "remembered" in his extreme old age. They were and are unbothered by the complete absence of any objective link between Sollas and Piltdown, by the vacuum regarding "evidence," or by the fact that the Sherborne horse is still today the subject of scientific dispute and may not have been a hoax at all. They offer no answer to the question of why Sollas never disclosed the truth of his elaborate hoax, the disclosure which was supposed to supply the rousing denouement. There—since nothing of any importance remains to be said—the eerie episode of accusations between dead men may be dropped.

The notably poor quality of the cases brought against Butterfield, Abbott, and Sollas, the complete lack in them of grasp as to both details and the general picture, is continued with the next two figures: Grafton Elliot Smith (the Piltdown neurologist) and Martin Hinton (then a young assistant in zoology at the Natural History Museum). Any extended rehearsal of the accusations against these two, calling up the same sort of directionless "evidence," would amount to tedious repetition of points already made. Concerning neither man does there exist more than the weakest conceivable set of circumstances. In fact, the case adjudged against Smith shows how very little it takes in the Piltdown jurisdiction to bring a serious indictment. On the same flabby basis it would be no trick at all to involve a dozen others still untapped—Louis Salzman, for instance, or Harry Morris or Captain St. Barbe himself, and almost any name on the roles of the Sussex Archaeological Society.

The Smith case was first proposed in 1972 in the book *The Piltdown Men* by Ronald Millar. But so convoluted is it, and so muddily described, that it defies reduction to a neat synopsis. Blended in a mix of "suspicious" circumstances are such matters as Smith's failure to guide Woodward and Keith in an accurate reconstruction of the Piltdown skull, his lengthy residence in Egypt where he worked with many old skeletons, his link with an Australian fossil, the Talgai skull, which was briefly invoked by a few to support Piltdown (but which has no part in the actual story). There was also Smith's presence in England on separate visits in 1907 and 1908, when he supposedly planted the fake Piltdown skull in the pit (sidestepping the fact that he was not based in England during those years).

The motive for the hoax ascribed to Smith by Millar is the convenient one of overweening ambition, coupled with Smith's desire to vindicate his own long-held theories on human evolution—all combined, somehow and for some obscure reason, with a manic wish to embarrass his colleagues. What Millar further ventures along these lines, the question of motive, inadvertently supplies a revealing admission, curiously naive, as to his own competence in the matter. The passage below may stand as a prime illustration of the lax attitude, especially in recognizing what constitutes evidence, that now pervades Piltdown literature:

> I was first inclined toward the second motive. I thought that as Smith grew in professional stature, and the forgery refused to let itself be

discovered, he allowed the matter to stagnate. He could not do otherwise, for a sudden revelation at this late stage would be highly suspicious. But as my research advanced, and I realized that Smith was a highly likely suspect, my view of the actual motive changed. In its place came the conviction that Smith would have loved a chuckle at the expense of what he thought, possibly correctly, was stick-in-the-mud palaeontology and anatomy. Somehow the whole affair reeks of Smith.

The leisurely phrases *inclined toward* and *came the conviction*, along with the indulgent *somehow reeks of*, are not nearly sufficient to meet the case of a scientist who has legitimately earned permanent rank high on the list of actual accomplishment. But a page or so after that disturbing passage, it is pleasant to note, Millar proceeds to redeem himself by placing in the record a straightforward confession of the sort not often met with among Piltdown investigators. "Although the realization that Sir Grafton Elliot Smith," he writes, "might be the hoaxer dawned on me about halfway through the preliminary research for this book, try as I may I have not been able to come up with concrete evidence of the Australian's participation." The question of why, lacking some degree of solid evidence, the Smith indictment was pursued at all, let alone so avidly, in such detail, and with such apparent conviction, need not be pressed.

Young Martin Hinton, of the British Museum zoology department, but by then in retirement, might well have gotten away unscathed had he not called attention to himself in 1953, just following the Piltdown exposure, by writing a letter to the *Times*. He and some of his friends at the museum, he said, had always rejected the jaw as being that of a chimpanzee, though they never suspected a fraud. The truth, he thought, might have come out much sooner than it did if Woodward had only permitted unrestricted study of the original bones, rather than having people rely on the casts:

Had the investigators been permitted to handle the actual specimens, I think the spurious nature of the jaw would have been detected long ago. Apart from the time when the actual specimens were exhibited in a showcase in the Geological Gallery I got within handling distance only on one occasion, and then I did not handle them.

Eoanthropus is a result of departing from one of the great principles of palaeontology; each specimen or fragment must be regarded as a separate document and have its character read. Here that principle was abandoned and an unnatural association of diverse things was made on what was considered to be the probabilities.

In 1912 I was working as a volunteer in the Geological and Zoological departments. I did not see the Piltdown material until the reading of Smith Woodward's paper at the Geological Society. As soon as I saw the jaw, and later the canine tooth, I knew that had they come into my hands for description they would have been referred without hesitation to the chimpanzee which was already known to occur in some of the Pleistocene deposits of Europe . . .

Hinton's charge that the Piltdown bones were not made available for study by others is both true and false. They were often, on request, put at the disposal of qualified scholars, but only for brief periods of some hours or days. It is true, of course, that the casts made from the originals could not be expected to faithfully reproduce every tiniest feature of value or interest to be seen in the originals. Probably Hinton was right, and if the complete original set of specimens had somehow been openly and permanently under the eyes of many scholars, the reality would soon have become apparent. Why this reasonable approach was not urged *at the time* must in great part be laid to Woodward's influence. If there was anything further of significance to be seen in the originals, ran the assumption, then Woodward would see it.

Hinton died in 1961. Twenty years later a ten-part article on Piltdown in *New Scientist*—ignoring the total lack of evidence regarding Hinton, but nevertheless employing an omniscient viewpoint—involved him in a plot at Piltdown whose twists seem better fitted to the topsy-turvy world of Gilbert and Sullivan. Since then Hinton has become a convenient marker in the Piltdown parlor game, finding a conspicuous place in the several conspiracy theories, as prime mover, coconspirator, or lackey.

Only such terms as "unbridled" and "labyrinthine" are adequate to describe the wildly speculative case brought in 1986 against Samuel Woodhead (died 1943) and John Hewitt (died 1954), who are portrayed as coconspirators. While vaguely and partially rumored earlier, the case against these two was first proposed in detail in two articles by Peter Costello in the prominent journal *Antiquity*. Its headlong mixture of supposition and assertion is based mainly on what can only be called a surprisingly reckless use of hearsay.

Samuel Woodhead is the Uckfield chemist who was mentioned in the official 1912 Piltdown report as having analyzed part of the skull. He was also mentioned in Dawson's 1913 article as having helped in an early and fruitless one-day search of the pit before Dawson's approach to Woodward. Scattered references show him to have dug at the pit on a few other occasions, once finding part of a badger's tooth. It was some unexpected and belated reminiscences by Woodhead's two sons—memories definitely confused and misleading—that led to his being put in the dock, to the sons' utter consternation.

John Hewitt, another chemist, first touched the lives of Woodhead and Dawson in 1898—and Dawson's *only* then—when he rendered a dissenting judgment on a discovery of natural gas in Sussex. Woodhead and Dawson disagreed with his opinion, and it seems were eventually able to prove him wrong. It was a decade later, avers Costello, that the still-simmering Hewitt, now with the coerced Woodhead on *his* side, perpetrated the Piltdown hoax as a means of venting his old anger against Dawson. (There is no effort to make this odd alignment, or the delayed reaction, appear reasonable.)

But that is not all. From there, Costello goes on to sketch out a conspiracy that drew into its tangled web a veritable horde of London's elite. Involved are any number of leading British scientists, named and unnamed, at the Natural History Museum, and at Oxford and Cambridge universities. Some of them are new to the Piltdown inquiry, while others are chosen from the growing list of established suspects.

Every paragraph in Costello's two articles, or almost, harbors single-source statements, charges, and claims that are patently exaggerated, unreliable, or provably false. Yet each is soberly and formally presented as a firm link in the chain of evidence. One example, fundamental in its implications, must suffice.

It was some time after Dawson came into possession of the first piece of the skull, writes Costello, that he "found more pieces" (plural). These pieces Dawson showed to Woodhead, his longtime friend, who promptly "tested the pieces and found their composition was the same." Those very same "tests," adds Costello, are specified in the 1912 report. True, Woodhead and testing are mentioned in the report, but the pertinent passage says something quite different from what Costello describes. In fact, concealed within the passage is a record of one of the forger's most signal triumphs. A single paragraph in length, it reads in its entirety:

A small fragment of the skull has been weighed and tested by Mr. S. A. Woodhead, M.SC., F.I.C., Public Analyst for East Sussex and Hove, and Agricultural Analyst for East Sussex. He reports that the specific gravity of the bone (powdered) is 2.115 (water at 5 C. as standard). No gelatine or organic matter is present. There is a large proportion of phosphates (originally present in the bone) and a considerable proportion of iron. Silica is absent.

Not "pieces," but only a "small fragment" was tested by Woodhead, a fragment not specifically identified in the report, but belonging to the cranium. No part of the jaw was analyzed by Woodhead at the time, or by anyone else. If at the start the jaw *had* been properly tested by a chemist, its higher organic content would have sharply marked it off from the cranium, rendering still more obvious its lack of relationship with the human skullcap. Why it was that not one of the many experts on both sides of the debate thought to raise this point is perhaps Piltdown's most enduring puzzle. Part of the answer certainly is that very paragraph, with its reassuring chemical references. Subtly it implies that the other specimens, including the jaw especially, would at some time be similarly tested.

Disturbing enough is the woefully flawed accusation brought so blithely against Woodhead and Hewitt. The fact that it appeared in the respected pages of *Antiquity* only deepens the surprise. More astonishing yet is what the journal's editor, Glyn Daniel, a respected archeologist in his own right, wrote in introducing Costello's theory. The subject of Piltdown, Daniel stated, had been "ventilated many times in the last twenty years in our pages." Many candidates for the role of forger "have been put up and knocked down," until everybody had begun to wonder "whether we should ever know the truth." With that, unexpectedly, he added his personal endorsement: "Now we think we do, owing to the most careful researches of Peter Costello . . . coming to what we believe is the proper and final solution."

It was neither the first nor the last time that sheer giddiness about Piltdown overtook some scholar otherwise known to be thoroughly sensible and sound.

In the forty years that have passed since publication of Joseph Weiner's *The Piltdown Forgery*, counting books, parts of books, and articles in English alone, the subject has been treated on more than one hundred occasions. Of those hundred treatments, at least half venture serious discussions of the identity of the fraud's author, and here is encountered what must be labeled an egregious failure of technical—rhetorical—nicety. Far too often these writers on Piltdown offend in the slippery matter of probability-based argument.

Never sufficiently appreciated is the consistent overuse in Piltdown studies of the probability factor. Of course things that are *possible* and *probable* do have a legitimate function in arguments of all sorts, helping to bridge lacunae that often occur in a developing line of thought, factual gaps that would otherwise forestall investigation. But probability is a delicate tool indeed, and must always be wielded with a proper measure of restraint. What is *possible* must be closely interwoven with what *is*—hard fact—as its backbone. Allowed to float on a cloud of pure surmise, invoked too frequently—the temptation for this is strong—it ends by dissipating its effect, defeating itself, though not every reader is alert to that unintended result.

By virtue of this eminently flexible instrument, anyone who knew Dawson, or who spoke with him more than twice, today is in real danger of standing accused as his accomplice, if not as prime mover. All that is required to build this type of case is a generous sprinkling of such phrases as *appear to have—perhaps—what could be more likely—doubtless—is it possible—it seems highly likely—one wonders if—may have been—had more reason than most—might well have been—if so, then—was it because.*

The final half-dozen phrases in that group have not been selected at random. They are all taken from a recent assault on still another suspect, this time the preparator, or model maker, at the Natural History Museum, Frank Barlow. In 1990, in the pages of *New Scientist*, the curator of odontology at the Royal College of Surgeons, Caroline Grigson, leveled an accusing finger at Barlow (died 1951) as having been Dawson's scientific mainstay in the fraud. She did so in the full realization that there existed no slightest bit of actual evidence to implicate him. Here are her musings upon a rather startling series of possibilities, in this short passage numbering no less than *ten:*

> Is it possible that Barlow, such an expert modeller, treated much of the forged material after it arrived at the museum, and made casts of the forged fragments, without noticing anything amiss? . . . Anyone who

has worked in a large museum knows that there are often odd bits of undescribed and unrecorded material lying around in cupboards. Who was more able to lay hands on such material than the department's preparator, F. O. Barlow?

What of a possible relationship between Barlow and Dawson? We know that Dawson had been in and out of the museum for many years, bringing in fossil fish and reptiles that he had discovered, and that in the course of doing so he became friendly with Woodward. It seems highly likely that he got to know Barlow at the same time. Certainly he would have known of Barlow's existence and his reputation, and he might well have sought out Barlow when he needed advice on the preparation of these fossils, and perhaps also when it came to manufacturing a few fossils. There were few people from whom he could have sought more expert advice or help.

Frank Barlow must have known that something was amiss but he had more reason than most not to expose the hoaxers. I suspect that it went further than this and that he may have been the skilled accomplice that Dawson needed.

Having posthumously impugned a man, otherwise well respected, solely on the grounds of sheer surmise, nothing but guesswork, Grigson then proceeds to supply him with a motive, and rather a surprising one. Preparator Barlow, she points out, as supplier of the Piltdown casts, which were sold, not given, to interested scientists and institutions, "had a financial stake in the whole business . . . the price was two guineas for the skull and one guinea for the jaw." Eventually a professional fossil dealer, Damon's of Weymouth, took over distribution of the casts, selling them on consignment for their maker at the museum, Barlow. This, implies Grigson, was not done for the usual reason, that is, to make the casts more readily available for study by other scientists. It was done, she suggests, to set up a profit-making sideline, and *that* venal purpose was the real reason Barlow joined in the fraud.

If Frank Barlow was indeed the Piltdown forger, and if he did it for a monetary gain, then it must be said that as a financial schemer he was no wizard. For all his painstaking efforts as a plotter, stretching over several years, for the enormous personal risk he ran, he received quite a modest cash return, about enough for a week's holiday with his family at Brighton.

8

The Writer Accused

As the sun rose higher, flooding his bedroom, he spoke in a feeble voice that was almost a whisper: he said he did not want to die in bed. Gently they helped him to his feet, pulled a robe around his shoulders, and guided him to a large basket-chair near a window. With his tearful wife seated beside him, and his two sons and a daughter in the room, for an hour he sat gazing silently out over the Windlesham rose garden, in the distance the South Downs rising dimly. Then at eight-thirty he quietly closed his eyes and drew his last breath. It was July 7, 1930, and England's most celebrated popular author, Sir Arthur Conan Doyle, aged seventy-one, was dead of heart trouble.

On the day of Doyle's death more than a dozen years had passed since the announcement by Arthur Woodward of the important Sheffield Park finds, firmly establishing Piltdown's scientific credentials. During that period Doyle had made no slightest reference to Piltdown's new status, or to any other evolutionary topic. All his former interest in fossils and in the ever more pressing question of human evolution had evaporated. He had, in fact, turned his back on literally all the many subjects that had once fascinated his restless mind. For long prior to his death only one thing mattered: advancing the cause of spiritualism. To this pursuit throughout his final decade Doyle dedicated his time, his literary talents, and a good part of his fortune (some 250,000 pounds).

Lecturing tirelessly on behalf of what for him was a true "religion," during the twenties he traveled many thousands of miles to many countries, at the same time writing book after book in which he explained and defended spiritualism. In the process he grew into the movement's best-known advocate, virtually its high priest. The complete convert, he questioned nothing, rejected nothing, accepted all—the table rapping, the ghostly materialization, the spirit photography, the slate writing, the automatic writing, the ectoplasmic displays, the levitated bodies and rising furniture, the mysterious jangling of untouched bells and tambourines, the disembodied voices in the dark, including those of his dead mother, brother, son, and nephew.

Time after time in public Doyle staunchly rushed to the defense of spiritualism's every facet and manifestation, as well as all manner of related phenomena. It was his celebrated name and eager support that first brought attention to what became known as the Cottingley fairies, photographs made by two young girls in the north of England showing tiny winged sprites and gnomes cavorting with them in the woods. These small creatures, Doyle announced in *The Strand* magazine, afforded thrilling proof of the existence of an invisible world, one often suspected but never till then revealed. The Cottingley photographs, he declared to an astonished public, "will mark an epoch in human thought." The book he rapidly produced on the event, boldly entitled *The Coming of the Fairies*, offered not only the original pictures, but a wealth of information on fairy customs, appearance, and population statistics. One whole chapter gave a report by a clairvoyant friend of his who after a visit to the Cottingley woods soberly listed clear sightings of water nymphs, wood elves, gnomes, brownies, and goblins.

Long before he died, the renowned creator of the great detective Sherlock Holmes, that inexorable reasoning machine whose instinct for the truth was unerring, had become a profound enigma, destined while he lived to perplex his admirers, and in later decades to bedevil a parade of biographers. Then, just over a half-century after his death, another startling development arose to confuse and confound his worldwide following. The mysterious Piltdown forger, came the charge, was none other than Conan Doyle himself. The accusation was no passing whim. It appeared in an official publication of the American Association for the Advancement of Science.

The lengthy article in which the charge was made reached a bemused public in the pages of *Science 83*, published in the fall of 1983.
Called "The Perpetrator at Piltdown," its author was an obscure academic named John Winslow, and its subtitle announced that a new and
more searching look at the famous fraud "turns up a surprising suspect,
the creator of Sherlock Holmes." Skillfully marshaling the evidence,
which appeared both extensive and varied, Winslow opened his case
against Doyle:

> ... there was another interested figure who haunted the Piltdown site
> during the excavation, a doctor who knew human anatomy and chem-
> istry, someone interested in geology and archaeology, and an avid col-
> lector of fossils. He was a man who loved hoaxes, adventure, danger;
> a writer gifted at manipulating complex plots; and perhaps most im-
> portant of all one who bore a grudge against the British science estab-
> lishment ... Sir Arthur Conan Doyle.

Especially from the Doyle camp, the reaction to Winslow's article
was swift and decided, amounting to a summary dismissal of the charge
as irresponsible and even "malicious." A man of Doyle's public stature
and known integrity, all insisted, was far above such an underhanded
escapade. The evidence was flatly rejected as wholly circumstantial, not
worthy to be taken seriously. Sherlock Holmes himself was promptly
quoted in Doyle's defense, a cogent remark of his made in *The Boscombe
Valley Mystery:* "Circumstantial evidence is a very tricky thing. It may
seem to point very straight to one thing, but if you shift your own point
of view a little, you may find it pointing in an equally uncompromising
manner to something entirely different."

A harsher view of the widely read article, and of the motives behind
it, was taken by some defenders. One incensed commentator branded it
as a flagrant hoax in itself, put forward as a "cheap publicity hype by
Winslow, in collusion with the AAAS." Its object was obviously "to
boost circulation and perhaps pull the magazine out of the red. *Science
83* had been losing money steadily for three years ... The schemers
delivered reprints of the article to the news media weeks in advance."

Many others, less concerned, shrugged it off with slight comment.
Stephen Jay Gould, Harvard anthropologist, saw it as "an evidence-free
argument" resting on nothing more than speculation. Ian Langham, a
professional paleontologist in Australia who was thoroughly familiar

with the Piltdown story, brushed it aside as not only "ill-supported" but also—and infinitely more serious—"obfuscatory of science itself." In 1988 in Canada a thirty-six-page booklet appeared entitled *The Curious Incident of the Missing Link*, written by Douglas Elliott. After lengthily analyzing the evidence against Doyle, it concluded that there was in fact no case at all: "Evidence has to be more than just suggestive, and the logic that ties the evidence together must be incontrovertible. It is particularly odious when shoddy techniques are used to malign the character of someone who is incapable of defense."

Doyle's only surviving child, his daughter Dame Jean Conan Doyle, some days after the appearance of the offending article responded with an indignant if controlled statement sent directly to *Science 83*. The incredible charge against her father, she insisted, "is simply not true," and the article itself nothing but "a piece of silly season journalism, a mere frivolity." The Piltdown imposture, she declared, "would have been quite contrary to Doyle's moral nature," and in any case in 1912, the year of Piltdown's discovery, her energetic father had been much too taken up by a wide variety of demanding tasks, including public duties in addition to his continuous writing. The unscrupulous Winslow, she finished, had taken a noble-hearted man revered by readers the world over and turned him into a "shabby little trickster." (*Science 83* did not print the letter, and it found its way into print only in 1987 in *The Baker Street Miscellanea*.)

The agitated response evoked by Winslow's thesis, a predictable combination of indignation and indifference, was understandable. No man could have seemed less likely to have perpetrated an elaborate fraud than this famed, hugely successful, and beloved writer. While he lived Doyle enjoyed a lofty reputation with the public for integrity and personal honor (not impaired even by his ties with spiritualism, though his intelligence and common sense were indeed seriously doubted). Even the mere mention of his possible culpability in the Piltdown affair struck many people as verging on the ridiculous. Yet it must at last be frankly stated that this dismissive reaction to the charge is decidedly out of place, quite simply wrong. While it may be largely circumstantial, the case against Doyle is by no means tenuous, nor does it lack elements that require and even demand sober review. The possibility that the creator of Sherlock Holmes was also the creator of Piltdown Man may not be passed off lightly.

In some ways, the brief offered by Winslow is a fairly clumsy effort, often failing to take full advantage of the materials at hand. (This fault, largely literary, may perhaps be laid to an editor of the magazine, Alfred Meyer, who is given credit as coauthor.) In setting forth Winslow's various arguments below, an effort is made to present them in their strongest light, particularly by including several items previously overlooked. Only by making against Doyle as solid a case as may be assembled can its merits be fairly judged.

The ancient town of Crowborough in Sussex sits some twenty miles south of London, and just to the north of Uckfield. On its outskirts is Windlesham, the large house that for more than two decades, beginning in 1907, was the home of Arthur Conan Doyle and his family (still standing, it is now a nursing home). It was at Windlesham that there occurred Doyle's first direct, traceable contact with the Piltdown story when in the fall of 1911 he welcomed Charles Dawson as a guest.

Actually the two men may have met earlier, following a May 1909 inquiry by Doyle to the Natural History Museum in London about some dinosaur footprints found near Windlesham. Arthur Woodward had responded to that inquiry by mail, also asking Dawson to look in on Doyle, an assignment that Dawson may or may not have fulfilled. (About this time Doyle wrote his mother: "I have another expert of the British Museum coming on Monday to advise me about the fossils we get from the quarry opposite. Huge lizard's tracks.") Some time afterward—no later than the spring of 1911—Woodward also went down to Windlesham, where he discussed dinosaurs with the writer, but of that visit there is no separate record, only a passing reference in a letter. Dawson's November 1911 visit thus marks the first focused contact between Doyle and any Piltdown figure.

Fortunately, Dawson on returning to his home in Lewes the evening of the visit took the trouble to write Woodward about the day's events:

My wife and I went to lunch with the Conan Doyles at Crowborough today and to see the great fossil! I regret to say it was a mere concretion of iron and sand. Sir Conan and the ladies pointed out several "striking resemblances" to the "carcasses" of various animals, all ultimately destructive!

But the visit was not altogether lost for as I was trying to draw Sir Conan away from the hope of finding much in the sandstone and directing his attention to the drift deposits above I espied a beautiful flint arrowhead embedded, and in view of us all. Subsequently we found washed flints; and so I started him off on a new and I hope more fruitful enterprise.

I was so sorry at his disappointment—he is such a good fellow—but the neolithic find revived him a lot. Of course, I have given him the arrowhead. Do not trouble to answer this. I hope to get something better worth your attention some day.

Unless Dawson paid a subsequent, unrecorded visit to Crowborough, it was on this November 1911 visit that he heard about the new novel, *The Lost World*. Doyle was just then completing the manuscript, with the story scheduled to begin running as a serial in *The Strand* in April 1912, book publication to follow in the fall. Dawson's later description of the plot, brief as it was, shows that he was told a good portion of the book's action. "A sort of Jules Verne book," he described it to Woodward six weeks before the *Strand* serialization began, "on some wonderful plateau in S. America with a lake, which somehow got isolated from Oolitic times," and where ancient animals and vegetation still survived. There is, however, one tantalizing omission in Dawson's obviously condensed description: nothing is said of the fierce tribe of ape-men, their bodies covered in red hair like orangutans, which in Doyle's story plays a leading role.

That Doyle was a visitor to the Piltdown pit, as Winslow claims, is true, though how often and just when the visits took place are open questions. Doyle's daughter, apparently after consulting her records (she cites no source), states that there was only a single visit, made on December 16, 1912, by auto. But earlier testimony obtained by the original investigators indicates that in 1912 Doyle was "one visitor who came back two or three times," and with his admitted interest in all that was happening at the site there is little reason to doubt this. Whether this can be taken to mean, as Winslow has it, that Doyle "haunted" the excavations, seems doubtful, though on that critical point there is no way now to be sure. Surreptitious visits were always a possibility, and on this likelihood Winslow relies heavily:

Doyle was also a prodigious walker who thought nothing of setting forth on long jaunts, "geologizing" as he went along. There can be little doubt that he often visited the relatively unguarded site, either by walking up the driveway that passed next to it or by peering over the hedge to observe the progress of the excavations.

Peering from behind the hedge, Winslow apparently means, at times when no others were to be seen at the pit (it can hardly mean peering over the hedge at the diggers as they worked). Conceivably this is meant to include clandestine invasions of Barkham Manor under cover of darkness.

When to all the above Winslow adds the reminder that the robust Doyle, then in his early fifties, lived only some eight miles from Barkham Manor, was a medical doctor possessing an intimate knowledge of chemistry and human anatomy, and also had a strong interest in prehistory and paleontology, he has earned the right to a respectful hearing. On such preliminary indications, any competent police investigator would be inclined to look closer.

Having shown that Doyle as forger would have had both opportunity and competence, Winslow addresses next a more searching question. "Where could the hoaxer," he asks, "have obtained the jaw of an orangutan who lived somewhere in the East Indies and died approximately in the 12th or 13th centuries?" Surprisingly, he is able to answer that Doyle might have had little or no trouble getting his hands on such an unusual specimen, for two of his close friends had direct access to rare fossils.

The first was Cecil Wray, a British magistrate who had served in the Malay Peninsula and was a fellow of the Royal Anthropological Society. In addition, Wray's brother was head of a Malayan museum in which was displayed a large collection of animal fossils from Borneo, home of the orangutan. The second friend was the well-known American phrenologist, Jessie Fowler, who in connection with her calling owned an impressive collection of human and animal skulls. Some of these from time to time she would sell to interested buyers, and during the pertinent years she was often in London. At least once she stayed as a guest at Windlesham, where she gave Doyle a phrenological reading.

Concerning the animal or mammalian fossils dug from the Piltdown pit, Winslow's unveiling of the evidence, despite its provisional nature, becomes more than slightly arresting:

Several of the fossil mammal remains that were salted into the Piltdown gravel pit have since been identified as coming from the

Mediterranean. The likely sites include Malta and a fossil cache in the Ichkeul area of Tunisia, which was not known to palaeontologists until 1946. Whoever was the hoaxer had to have access to such exotic materials. In 1907, some two years before any fossils were discovered at Piltdown, Doyle visited archaeologist Joseph Whitaker, one of the few scientists who had frequently been to the Ichkeul region.

A few months after that meeting, Doyle and his bride—it was his second marriage—honeymooned for two months in the eastern Mediterranean. In all probability they went ashore at Malta, a British port, in late November or early December on their return voyage. Coincidentally, the *Malta Daily Chronicle* announced on November 16 the discovery of the fossilized remains of a hippopotamus by a workman excavating a limestone fissure on the island. One of the planted items at Piltdown was a hippopotamus tooth whose form and chemical content indicate it came from a limestone chamber on one of the Mediterranean islands, Malta being regarded as most likely.

Two years later, in 1909, before the discovery of any of the Mediterranean fossils at Piltdown, Doyle and his wife cruised the western Mediterranean. They visited Algeria and almost certainly Tunisia. Shortly after, Doyle wrote a story about Carthage, located not far from Ichkeul. The ship also stopped at Malta and Corsica, another possible source for the hippopotamus tooth. None of the other individuals suspected of being the hoaxer is known to have visited these islands or Tunisia. The timing of Doyle's travels was perfect . . .

Finally, there are the other fossil mammal fragments, including beaver teeth, that were salted into Piltdown but most likely came from Norfolk or Suffolk. For years prior to Piltdown, Doyle vacationed in Norfolk near towns that boasted excellent golf courses, as golf was one of his favorite games. These included the Sherringham Golf Course which abutted the East Runton deposit, a confusing collection of fossils ranging from late Pleistocene beaver to a variety of early Pleistocene animals, the same unusual kind of mix found in the Piltdown gravel pit.

Also handily accounted for by Winslow are the flint implements associated with Piltdown, so valuable for dating the skull. One type in particular, those with a white cortex, can be linked to a small desert town in Tunisia called Gafsa, which was the site of "the largest palaeolithic flint factory in North Africa." Thus Doyle during his travels could have collected the flints on his own, or he could have obtained them from still another friend, the distinguished author Norman Douglas. An avid collector of ancient flints, with hundreds of examples in his possession, Douglas in 1910 had spent several months rummaging for flints in Gafsa.

Winslow had no need to insist on Doyle's abiding interest in matters prehistoric. But to show that at least his familiarity with such things predated Piltdown by many years, he quotes a passage from the most popular of the Holmes tales, *The Hound of the Baskervilles*, published in 1901. It recounts the initial meeting in Baker Street between Holmes and Dr. Mortimer, who promptly ventures a personal observation:

> You interest me very much, Mr. Holmes. I had hardly expected so dolichocephalic a skull or such well-marked supra-orbital development. Would you have any objection to my running my finger along your parietal fissure? A cast of your skull, sir, until the original is available, would be an adornment to any anthropological museum. It is not my intention to be fulsome, but I confess that I covet your skull.

Overlooked by Winslow in the same novel are several other passages touching on the subject of anthropology. Between Holmes and Dr. Mortimer, for instance, there is a second exchange which, though brief, reinforces its author's easy acquaintance with the abstruse topic of skull morphology. Holmes speaks first:

> "I presume, doctor, that you could tell the skull of a negro from that of an Esquimaux?"
> "Most certainly."
> "But how?"
> "Because that is my special hobby. The differences are obvious. The supra-orbital crest, the facial angle, the maxillary curve, the—"

Later in the story Watson in Devonshire reports to Holmes that Dr. Mortimer has been doing some excavating and has dug up "a prehistoric skull which fills him with great joy." Watson himself, while walking the moors, encounters the cave dwellings of "prehistoric people," and he reflects that he would not be surprised to see "a skin-clad, hairy man crawl out from the low door, putting a flint-tipped arrow on to the string of his bow."

Still other passages in Doyle's writings with at least a tentative bearing on Piltdown are also ignored by Winslow. In *The Three Garridebs*, a Holmes story written before 1916 but not published until almost ten years later, there is pictured an unusual parlor in a London house meant to provide the setting for a classic Holmes denouement. Here Piltdown really should have found a place, and by name, but it is conspicuously absent:

> The room was as curious as its occupant. It looked like a small museum. It was both broad and deep, with cupboards and cabinets all around, crowded with specimens, geological and anatomical . . . Here was a case of ancient coins. There was a cabinet of flint instruments. Behind his central table was a large cupboard of fossil bones. Above was a line of plaster skulls with such names as 'Neanderthal,' 'Heidelberg,' 'Cro-Magnon' printed beneath them . . .

Citing as examples of the cabinet's contents three of the fossil men then known to science, for no clear reason Doyle passes over the very one that would have been most readily recognized by his English readers.

In another of Doyle's adventure novels, *The Land of Mist* (not connected with Holmes), there is a spiritualistic séance at which a veritable ape-man is materialized to stalk in all its bestial glory around a darkened chamber. But again the creature is presented as a pithecanthropus, not the much more famous Piltdown. The same oversight occurs a third time, in *The Lost World*, where Doyle twice refers to pithecanthropus but never to Piltdown. (News of Piltdown, even advance news or rumors, may have been too late for insertion in the *Strand* serialization, but apparently it would have been in good time for book publication late that fall.) Why Doyle should have steadily avoided naming the English ape-man in his stories is a question worth pondering. It does not seem accidental.

Another Sherlock Holmes tale, written during this period, *The Devil's Foot* (published in *The Strand* in December 1910), has a passing reference which shows that Holmes' vast stores of knowledge included a firm grasp of anthropology. Investigating a murder in Cornwall, the detective finds himself momentarily stumped, upon which he calls a halt to the inquiry. "It won't do, Watson!" says Holmes with a laugh. "Let us walk along the cliffs together and search for flint arrows. We are more likely to find them than clues to this problem." As the two stroll together along the cliff edge, Holmes lays out a variety of possible clues for his duller companion. Then, recalling his purpose in visiting the cliffs, he suggests that they put aside the case until more information is obtained, "and devote the rest of our morning to the pursuit of neolithic man." At this, Watson informs the reader, "I have often commented upon my friend's power of mental detachment, but never have I wondered at it more than that spring morning in Cornwall when for two hours he discoursed upon Celts, arrowheads, and shards, as lightly as if no sinister mystery was waiting for his solution." A two-hour discourse, presumably, would require much more than a dilettante's acquaintance with the subject. Of course it is a mere trick of the writer's art to endow a character convincingly with great stores of knowledge, but taken together, all these passages foster the impression that the author himself did in fact possess such knowledge.

It is with Doyle's novel *The Lost World* that Winslow succeeds in making perhaps his strongest impression. Admittedly the sheer coincidence in timing between book and fossil discovery does seem wonderful. Here was that rare thing in literature, a fictional plot that was entirely original, treating in colorful detail the supposed survival into modern times of a long-extinct portion of the planet's history (before Doyle no fiction writer had touched the theme). Yet the book made its appearance at the very moment that a highly dramatic aspect of that history suddenly showed up in real life, the two entities coming to birth not only simultaneously but in close physical proximity. In addition, both immediately drew wide attention, Piltdown necessarily much more than the novel (though the novel was warmly welcomed, was quickly published in the United States, and was translated into several Continental languages).

Even the book's main characters, says Winslow, contribute to the uncanny coincidence, and it is true that for at least one of the characters the suggestion seems well founded. The figure of Professor Summerlee, described as an expert in the field of comparative anatomy, seems almost transparently a portrait in miniature of Arthur Smith Woodward, down to the superior attitude and Vandyke beard. "A tall, thin, bitter man with the withered aspect of a theologian," Doyle sketches Summerlee, who exhibits throughout the story "a dry, half-sarcastic, wholly unsympathetic manner," his aloofness set off by a "thin, goatlike beard." If in fact Doyle based Summerlee on Woodward, then it appears he must have known the scientist more intimately than has been thought, and must have met him more often than is indicated by Woodward's one known visit to Windlesham, in 1911. Unless it was concealed, some small sign of this relationship should be traceable.

Winslow's trump card in his study of the novel comes with his description of the "lost plateau" in South America on which the dinosaurs and ape-men are found. At one point the tale's narrator manages to climb to the top of a very tall tree, from which he draws a map of the whole region in a notebook. The landscape he viewed below was "of an oval contour, with a breadth of about thirty miles and a width of twenty. Its general shape was that of a shallow funnel, all the sides sloping down to a considerable lake in the centre." The lake is about ten miles in circumference and is surrounded by woodlands. "I could see at my feet the glade of the iguanodons, and further off was a round opening in the trees which marked the swamp of the pterodactyls."

The resulting map, roughly sketched, accompanies the text. On it are marked a number of added features, such as a "pit for dinosaurs" and the field of the culminating battle with the tribe of ape-men. This map Winslow promptly and happily identifies as exactly corresponding to the Sussex countryside, specifically the Weald, its center being the very location of Piltdown itself. The terrain's fictional features Winslow is able to pair with actual locations in Sussex, and he insists that this "intriguing resemblance" constitutes still another "incriminating link" between Doyle and the hoax. He gives his own careful sketch of the Sussex landscape, placing it alongside Doyle's map, and the two indeed seem to bear a striking similarity.

Here, certainly, is the point at which to quote an incidental remark made in the novel by one of its minor characters, a remark understandably highlighted by Winslow. In London two men are discussing a bone and a photograph that have been claimed as evidence for dinosaur survival in South America. One of the two is an unbeliever, and he observes offhandedly that the so-called evidence is probably spurious and must have been "vamped up for the occasion. If you are clever and know your business you can fake a bone as easily as you can fake a photograph." Juxtaposing that frank assertion with the actual fakery then going on a few miles to the south of Doyle's writing desk at Windlesham must give a jar to any reasonable mind.

Coming to the pivotal question of motive, Winslow treats it rather deftly, suggesting it arose from Doyle's fervent, even fanatical support of the spiritualist movement. England at the time had many outspoken opponents of spiritualism, and one of the staunchest was Ray Lankester, friend of Woodward and Keith. Repeatedly Lankester urged that aggressive tactics should be employed in combating what he saw as a criminal imposition on the unwary. In fact, he had himself taken action against an American medium, Henry Slade, who years before had been the rage of London. Winslow explains:

> Lankester had arranged to attend a seance with Slade, the purpose of which was to communicate with a spirit. The spirit would manifest itself by writing on a blank slate . . . Suspecting Slade of having tampered with the slate, Lankester decided to "test my hypothesis . . . by conducting a crucial experiment." After the slate was presented as clean, but before he heard any noise of writing, Lankester snatched it and found a message on its surface. This Lankester viewed as certain evidence of an intent to cheat and defraud him of the fee he had paid to Slade, evidence that "would be convincing to persons not already lost to reason." A magistrate agreed, though Slade was released on a technicality and left England as expeditiously as he could.
>
> In a single stroke, Lankester had become the Spiritualists' *bete noir.*

Doyle's first reference to the Slade-Lankester affair, Winslow correctly states, occurs in a short story, "The Captain of the Pole-Star," published in 1883. Even if mediums such as Slade *were* guilty of fraud, Doyle implies in the story, spiritualism must not be condemned for that reason alone. "But Lankester had used just this kind of reasoning," says Winslow. "Piltdown would provide a chance to reverse the tables by applying the same kind of logic: if science swallowed a scientific fraud like Piltdown Man, then all of science, especially the destructive and arrogant evolutionists, whom Doyle called the Materialists, could be condemned." Subsequently, Winslow adds, Doyle made frequent reference to "his adversaries, the Materialists," singling out Lankester as one of the most flagrant examples.

Here was a profoundly personal reason for the hoax, and strongly emotional. A specific "inviting target" had also been provided: Ray Lankester, the scientist most rabidly opposed to spiritualism.

If Doyle did have the motive Winslow imputes to him, then at some point *revealing* the imposture would be imperative, full disclosure being the scheme's necessary outcome, its sting. But it is just here that Winslow's argument is at its weakest. On several occasions, he suggests,

Doyle "may well have tried" to engineer a public unveiling, only to have his intention frustrated by the obtuseness of his victims:

> Doyle was a sportsman as well as a jokester. As he was an expert cricketer who had played on some of the country's top teams, what could be better than to place a cricket bat "in the hands" of Piltdown man? In 1914 a portion of a fossil elephant femur was discovered at the site. When it was formally described at a Geological Society meeting, a scientist rose to state that "he could not imagine any use for an instrument that looked like part of a cricket bat." He further believed in the possibility "of the bone having been found and whittled in recent times."
>
> But most of the scientists either ignored it or preferred to believe that the object was a genuine palaeolithic tool, though none could assign it a plausible function.
>
> The following year, 1915, another fossil deposit was discovered by Dawson one or two miles from the original site. Called Piltdown II, this site also may represent an attempt by Doyle to strain the credulity of scientists to the point where they would question the authenticity of both deposits . . .

Instead, Piltdown Two—the Sheffield Park finds—unexpectedly had precisely the opposite result of what Doyle supposedly had hoped. It actually bolstered the conviction of authenticity. "Such gullibility," Winslow remarks lamely in ending his lengthy presentation, "must have exasperated Doyle, or made him howl with laughter."

Some aspects of the Doyle case there is little use discussing. That he did or didn't, or might, or could have possessed the scientific competence to prepare the spurious artifacts, for example, is a question past the need of weighing. As a doctor, his grasp of anatomy and chemistry, obtained in medical school and put into practice for a decade before he turned full-time to authorship, would seem sufficient for purposes of the hoax. But if additional knowledge was needed, say in the fields of geology or paleontology, acquiring it would surely have been only a matter of application.

For procurement of the skull and the other materials, the same holds true. Even if it could be proved that on his two trips to the Mediterranean Doyle did not visit Malta or Tunisia, while in Norfolk did not play golf at the Sherringham Club, and could not have gotten what he needed from his friends, there would still have been other, quite adequate sources open to him. Most obviously, there were the professional fossil dealers in London—the firm of Gerrard's of Camden Town, for instance, or Butler's in Brompton—all of whom did a thriving business with England's museums, taxidermists, and private collectors. Gerrard's, a long-established company, offered "an enormous variety of specimens and bones," many of them of great rarity.

The supposed intimate link between Piltdown and Doyle's novel *The Lost World* proves to be far overstated. True enough, in general, is the parallel between the fictional plateau and the Sussex Weald. But Winslow has subtly stretched and distorted his own skimpy map of "The Weald" to make it more closely match the one given in the novel. Also, on his own map he plots only a few selected features, which promote a superficial identity but whose bearing on the case is not always evident.

Between the two patches of land, the plateau and the Weald, there are some significant differences not stressed by Winslow. At the center of the fictional plateau, for instance, is a large lake to which all the surrounding land slopes down. But this is a body of water and a funnel formation unknown to Sussex. Again, on his map of the Weald, Winslow draws an outline for a feature he names the "Hastings Sands." In form, size, and placement it closely images Doyle's plateau lake, and the result is an apparent visual link or clue. But these so-called "Sands" in reality are the Hastings Beds, an extensive layer of sedimentary rock underlying the central Weald. The vaguely oval shape Winslow gives to his "Sands" is quite loose, and in fact arbitrary. Also, the real Hastings Beds mostly lie far beneath the surface, and thus make no legitimate part of Sussex topography.

In any case, in the novel Doyle himself provides the first hint about his very generalized use of his home county for fiction. In his text he has a character who on inspecting a dinosaur footprint remarks, "I've seen them in the Wealden clay . . . they puzzled a worthy Sussex doctor some ninety years ago." The worthy doctor was real, one Gideon Mantell of Lewes, a pioneer of Sussex geology, and of course Doyle himself had been "puzzled" by iguanodon prints found near Windlesham. Another character in the novel describes the layout of the plateau and explains, "An area, as large perhaps as Sussex, has been lifted up *en bloc* with all its living contents." Winslow's map comparison carries weight only if the correspondence is extensive and precise, and if it covers something deliberately hidden. Neither of these requirements proves true.

The question must also be asked: if there *was* some close and secret correspondence between hoax and novel, what would it signify? Surely this is a fundamental point, yet here again Winslow's reply proves both incomplete and ineffectual. "The Piltdown hoax," he declares as if making a revelation, "was inspired by, or developed hand-in-hand with, the plot of *The Lost World* . . . the seeds of *The Lost World* appear to have been planted in Doyle's mind long before Piltdown was a site of recognized significance in anyone's mind." At that he simply drops the thought, saying not a word more about it, while leaving readers to wonder what he was getting at. In truth, it is hard to think of a convincing or even possible explanation of why Doyle, if he was the hoaxer, should have anticipated—or reflected or hinted at—the matter in the pages of a novel.

But perhaps the most telling observation to be made against any possible relevance of *The Lost World* to Piltdown relates to Doyle's picture of his imagined ape-men. It is correct that he covers them in reddish hair, suggesting orangutans, thus seeming to imply a link with the faked Piltdown jaw, later identified as coming from an orangutan. But the heads of Doyle's ape-men are depicted as quite unlike that of Piltdown Man. They actually contradict the then-prevailing belief which had produced the Piltdown form in the first place—that it was man's *brain* that led the way in the evolutionary march. When the novel's narrator climbs a tree to make his map of the plateau, on the way up he is startled by coming face to face with one of these creatures, the first to be encountered by any of the party. At a particularly thick and leafy branch he pauses to catch his breath:

I leaned my head around it in order to see what was beyond, and I nearly fell out of the tree in my surprise at what I saw.

A face was gazing into mine—at a distance of only a foot or two . . . It was a human face, or at least it was far more human than any monkey's that I have ever seen. It was long, whitish, and blotched with pimples, the nose flattened, and the lower jaw projecting, with bristles of coarse whiskers around the chin.

The eyes, which were under thick and heavy brows, were bestial and ferocious, and as it opened its mouth to snarl what sounded like a curse at me I observed that it had curved, sharp canine teeth.

Doyle's missing link, it can be seen, is far from being the large-brained Piltdown Man with its high, human forehead rising above an ape's jaw. Despite his saying it had a human face, Doyle's ape-man in both brain and muzzle is a veritable beast, an unevolved animal. Its lower jaw is "projecting," and those "bestial and ferocious eyes" peer from under brows that hang "thick and heavy," like those of any simian. This portrait of a creature supposedly half-man and half-ape has nothing markedly human about it. In writing his imaginary description, it may be concluded, Doyle never saw, never envisioned the form of skull that actually came from the Barkham Manor pit. Neither, it seems, did he know anything of the crucial straight canine found by Teilhard in that spoil heap. If he had, he would scarcely have specified a "curved" canine for his own creation.

The foregoing discussion was necessary and instructive, unavoidable if the charges against Doyle were to be treated fully and in detail. The result shows, at least, how questionable these charges can be made to appear by a patient review of the facts. But the true and proper point of departure for analysis of the Doyle-Piltdown matrix relates to a matter hardly touched on by Winslow: the question of Doyle's access to the pit itself. If Doyle is to be taken as a serious candidate for the role of forger, then it must first be shown that he was, or could have been, at the pit on certain specified days, or just before those days. This is because the planting of the bogus specimens was in reality an integral part of the proceedings.

The deliberate manner in which the various specimens of human and animal remains were introduced into the Piltdown gravel has never been well understood. Commentators on the case all assume that this operation involved little more than the surreptitious flinging down, almost at random, of bits of bone, animal teeth, and flints. Many writers on the case speak of the pit as having been "salted" once and for all as early as 1908, and Winslow in this regard is no different. His hurried, vague notion of the process is couched in much the same loose terms: "Since most of the remains were found on or near the surface, it required no great feat on the part of the hoaxer to insert them into recently exposed cuts or toss them into the spoil heaps where their discovery could be assured."

Now, even cursory study of the timing and the separate actions involved with the various individual finds will show that they could not have been "tossed" at random into or around the pit, and certainly not at the same time. All the many specimens must have been deliberately and carefully planted at different times, according to a set plan. Most importantly, the planting of each specimen must have been done just prior to its discovery. In some cases this would have been as little as an hour before or less, and in other cases perhaps several hours. But probably none of the finds had lain in the soil as much as a whole day (aside from the large bone implement, which would have been put into the ground up to several days before). In no other fashion could the hoaxer be sure that his precious materials—prepared in secret with such minute and painstaking effort—would be found and recognized rather than overlooked and lost forever.

The mere work of fabricating so many spurious items, each demanding infinitely delicate adjustment, must have cost the hoaxer many long hours of remarkably patient labor. Their intended discovery in the bewildering mix of dark, flint-strewn, pebbled soil in and around the pit at Barkham Manor could hardly have been left to the unpredictable course of ordinary excavation techniques. When any item was uncovered, it may be taken as certain that it had been deliberately put in the way of the particular finder. It follows that the planting of each separate item occurred that same day, and only shortly before it was found.

Just how that feat was accomplished for each find will be explained in its proper place further on. Here it is only necessary to acknowledge the fact itself: the several small pieces of cranium, the half-jaw, the canine, the many animal teeth, the bone implement, and the flints all were introduced into the gravel separately and at different times, over a period stretching from June 1912 to June 1914. (Missing from this list is any mention of Sheffield Park and Barcombe Mills. Those "finds," of course, were not "found" at all, and form a special aspect of the problem. They have no bearing on Doyle's possible involvement.)

It is in calculating the pattern and number of those repeated planting operations that the prodigious and complicated nature of the task unfolds. In 1912 discoveries were made on at least eight different days, and perhaps as many as ten or a dozen, all either on Saturdays or Sundays (a dearth of records for the mammalian and flint finds and one or two pieces of the cranium causes the indecision). The next year, the summer of 1913, there were fewer successful days, probably no more than five, again all on weekends. For 1914 the count can be fixed at

In Uckfield, East Sussex, 1912: Arthur Smith Woodward (*foreground*) with his son, Cyril, and Charles Dawson. The photo was taken soon after the start of the excavations at Piltdown.

The young priest Pierre Teilhard de Chardin, in England in 1912, soon after his ordination. He is standing on the shore near Hastings, site of the seminary, some miles south of Piltdown.

The Piltdown pit, 1913. Obviously posed, the photo shows Charles Dawson holding a sieve while Arthur Woodward inspects an object (ordinarily each would work separately). Digger Venus Hargreaves, instead of wielding his pick, stares at the camera.

The cranial fragments from the Piltdown pit. A total of nine small pieces turned up (in the photo several have been fitted together), along with various animal teeth and flint tools. (Not to scale.)

The bone implement (much reduced).

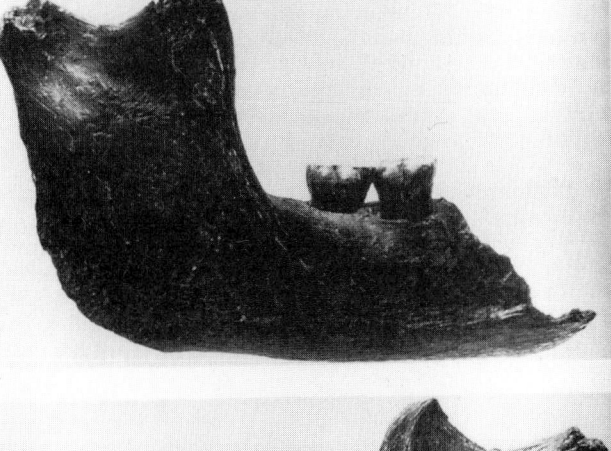

Found in the pit by Charles Dawson, the Piltdown jaw was for long the focus of controversy. The upper view shows its outer side; below, the inner. The two molar teeth revealed a human wear pattern.

Arthur Woodward's reconstruction of the "Dawn Man" skull, *Eoanthropus dawsoni*. It is based on the finds from Piltdown: nine cranial fragments, the jawbone, and one canine.

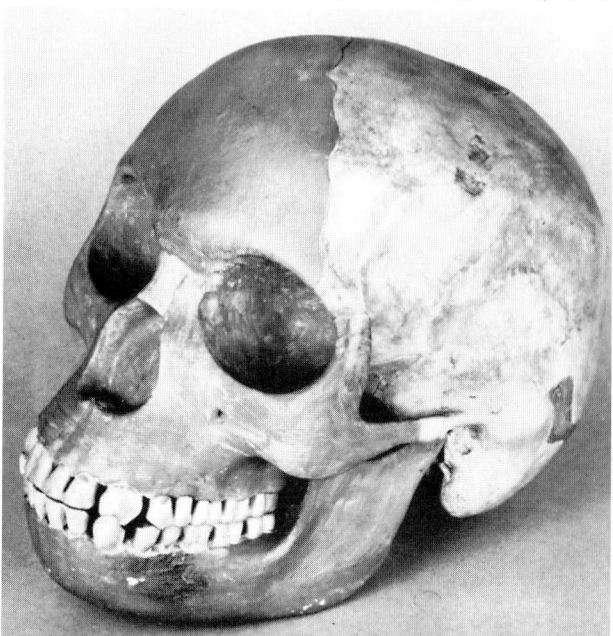

One of the conjectural "restorations" of Piltdown Man. This version, the work of Belgian scientist Louis Rutot, has the creature deftly using an eolith as a tool.

In July 1913, in the middle of the second season of digging, a large party of geologists descended on Piltdown by invitation. Some even invaded the pit itself, as is evident at the photo's extreme left.

In Lewes, Charles Dawson's residence enjoyed a prestigious location—a corner of his house, Castle Lodge, is seen to the left of the ancient castle gate. On the right is Barbican House, home of the Sussex Archaeological Society. (Photograph taken in 1993.)

In the basement of London's Natural History Museum in 1913, Charles Dawson (*left*) and Arthur Woodward examine the reconstructed Piltdown skull. As Keeper of Geology at the museum, Woodward had full charge of the remains.

Charles Dawson in ceremonial dress. A prominent lawyer, he was a fellow of two scientific bodies, but his great disappointment was his failure to win election to the Royal Society.

Participants in the original Piltdown controversy in a painting of 1915 by John Cooke. *Rear, from left:* Frank Barlow, Grafton Elliot Smith, Charles Dawson, Sir Arthur Smith Woodward. *Front:* Arthur Underwood, Sir Arthur Keith, William Pycraft, Edwin Lankester.

The Woodward family and a friend, with Charles Dawson, at the Piltdown pit, Easter 1914. From left rear: Dawson, Woodward, his daughter, Margaret (Mrs. Hodgson), Mrs. Woodward, and a friend.

Joseph Weiner, an anthropologist of Oxford, was born in South Africa, where he studied under Raymond Dart, discoverer of *Australopithecus.* In 1953 his efforts began the downfall of the long-established "Dawn Man."

Kenneth Oakley, a pale-ontologist with the British Museum of Natural History, became Weiner's close ally in the exposure. His revival of a neglected chemical test proved to be the key.

three, though four and even five is possible, and may include one or two weekdays.

Taken together, the Piltdown discovery days comprise at least fifteen, with as many as twenty-three not ruled out. Setting aside all else, if Doyle was the perpetrator, he must have paid stealthy visits to the pit on the Barkham Manor grounds—a situation open on three sides and in full view of the manor house—between fifteen and twenty-three times, all without being spotted. A few of those visits—to plant the mammal teeth and the flints—conceivably could have been made the night before, but most must be referred, at the earliest, to the very morning of the discoveries. (Work at the pit seldom began much before noon.) Also, as soon becomes evident, in a few instances the planting of the specimens took place *after* the day's work at the pit had begun, and while the digging was in progress. This fact alone precludes Doyle's involvement, as it does many of the others accused.

A final relevant fact now becomes obvious: for all these separate plantings, totaling at least fifteen, Doyle, if he was the hoaxer, would have needed specific advance knowledge about the excavators' schedule and intentions, as well as the party's makeup. Raising further complications for the hoaxer, during those first three years of digging at Piltdown there were any number of times when the party's plans were suddenly changed, or were uncertain until the last moment.

At the time of Piltdown, Conan Doyle was a busy writer in great public demand. He was also a much-traveled man and an activist, deeply involved during the period 1911–15 with any number of causes. (In 1912, among other things, he toiled mightily on behalf of the British Olympic Committee, and also gave much time and attention to the real-life murder case of Oscar Slater.) If the records of his movements in those years—letters, notes, diaries—could be consulted, it might well be decided whether he could have been in the Piltdown area on certain pivotal days. Such records do exist, but at present are closely held by the author's estate because of continuing family litigation (the thousand letters Doyle wrote to his mother, for instance, have yet to be released). The several biographies and books of reminiscence and commentary unfortunately are bare of the pertinent dates.

Coming to the question of motive, here again Winslow is answered with relative ease. Under a more minute inspection, the thesis so painstakingly built up by Doyle's accuser readily crumbles.

The Piltdown hoax, Winslow suggests, represented a sincere if misguided attempt by a committed spiritualist to repulse and expose the movement's "Materialist" foes. Its main target, supposedly, was the prominent scientist Ray Lankester, a man despised by Doyle for his public crusade against spiritualist mediums, and particularly for his destruction of the famous medium Slade. But in reply, a single uncontested fact may be cited that defeats such groping speculation. By 1912, and probably well before that, the situation between Doyle and Lankester had altered. No longer enemies, they had actually become good friends, and it was in the pages of Lankester's book *Extinct Animals*, published in 1906, that Doyle found much of the inspiration for his *Lost World* novel. In acknowledgment of this debt, he even offered the scientist a generous salute in his narrative. The main protagonist of *The Lost World*, Professor Challenger, is contradicted by a friend on the question of dinosaur survival:

> For answer the professor took a book down from a shelf. "This is an excellent monograph by my gifted friend, Ray Lankester!" said he. "There is an illustration here which would interest you. Ah! yes, here it is! The inscription beneath it runs: 'Probable appearance in life of the Jurassic dinosaur Stegosaurus. The hind leg alone is twice as tall as a full-grown man.' Well what do you think of that?"
>
> He handed me the open book. I started as I looked at the picture. In this reconstructed animal of a dead world there was certainly a very great resemblance to the sketch of the unknown artist.

Perhaps no further comment is needed, except to note that the picture of the Stegosaurus, with exactly that caption, may be found on page 208 of the Lankester volume.

But even if Lankester as an individual was not the target of a Doylean hoax, might not the fraud still have been spawned by the writer's fervid defense in general of a beleaguered spiritualism? Once more what appears a plausible supposition quickly buckles under the pressure of simple fact.

By his own admission—and it is supported by the available evidence—Doyle did not become a wholehearted, proselytizing spiritualist until the advent of World War One. Born a Catholic, he later turned to what he called "a broad Theism." An interest in spiritualistic phenomena accompanied the theism, but for a long time it was little more than the concern of a respectful student, admiring from a distance. As Doyle recalled more than once, it was quite definitely the year 1916, and largely the war's awful carnage, that precipitated his complete conversion to the spiritualist cause. Piltdown occurred nearly five years before that submission. It occurred perhaps ten years before Doyle conceived his strange, crusading notion that the ghosts and goblins to be found in every séance room and forest glade were meant to be the world's salvation.

Patient investigation of Winslow's charge stands firmly against any slightest possibility of Doyle's guilt. The creator of Sherlock Holmes was not the fabricator of Piltdown. Full corroboration for that verdict, final and convincing, comes with exposure of the true and undoubted culprit.

9

The Priest Accused

In mid-July 1912 at the port of Dover a tall, thin, narrow-faced young priest went aboard a channel steamer and sailed for France. He had been summoned by his superiors to continue his training as a Jesuit in his home country, and would not see England again for a year. With that departure, Teilhard de Chardin effectively removed himself as a suspect in the Piltdown affair, or so it would seem. If he was out of England he could not have been present at the pit during the remainder of the 1912 excavations to plant the subsequent finds.

Still, for four years Teilhard had lived and studied near Piltdown, was a budding paleontologist of some little experience, had been on more or less intimate terms with Charles Dawson, and had actually dug at the pit. On his return to Sussex late in the summer of 1913, he lost no time in rejoining the excavations and then promptly turned up the canine so necessary to the full success of *Eoanthropus.* All the special erudition required by the hoaxer, including a background in chemistry—he had taught the subject in a Jesuit high school before coming to England— he certainly possessed. No more than with Conan Doyle may the accusation against Teilhard be waved aside. Despite its unlikelihood, the charge that this now far-famed cleric had in his youth been implicated in a massive fraud calls for sober investigation.

It was inevitable, of course, and merely because he had been part of the excavating team, that from the very first the young Frenchman should have raised at least a dim suspicion in the minds of some. Though his name was seldom used, rather broad hints about his complicity began appearing in print soon after the initial exposure. Indeed with one leading anthropologist in the sixties, Louis Leakey, Teilhard's guilt became almost a personal obsession. In his many writings he made repeated, not-so-veiled references to Teilhard's guilt, and he actually planned and partially wrote an entire book on the topic. In 1971 Leakey was invited to speak, in his professional capacity, at a symposium reviewing Teilhard's scientific career, to be held at the University of California. He accepted, but duly warned the sponsors that he had "the very strongest belief that in his early and somewhat irresponsible days" Teilhard had taken "a considerable part" in the Piltdown forgery. Leakey spoke at the symposium, but the subject of Piltdown never came up, and he died in 1972 without completing his promised book. None of his evidence in the matter, whatever it may have been, has ever surfaced.

It was in 1980, somewhat suddenly and from an unexpected quarter, that there appeared what may be called the formal charge—not the usual glib surmise but an offering of hard evidence—against Teilhard as the Piltdown forger, or more properly, as an accomplice in the fraud. Though the dead Teilhard's adherents responded immediately and in some heat, the accuser, an American scientist of considerable repute and public standing, could not be ignored or shouted down. He was Stephen Jay Gould, Harvard anthropologist, professor of the history of science, and a popular author of scientific literature. His lengthy, detailed presentation of the case against Teilhard, a sober exposition based on much study, was carried in the magazine *Natural History*, an official publication of New York's American Museum of Natural History. Fifteen years later its reverberations still have not ceased.

One morning at breakfast in the fall of 1953, as he later recalled, twelve-year-old Stephen Jay Gould saw on page one of *The New York Times* the first newsbreak about the Piltdown exposure. He was just at the right age for "primal fascination," as he wrote afterward, adding, "My interest has never abated, and I have, over the years, asked many senior palaeontologists about Piltdown." Among all those who have tried to solve the Piltdown riddle, no other has had the subject in mind so seriously for so long. That reason alone makes Gould's pronouncement on the case worth hearing.

Few of the men he talked with over the years, says Gould, believed that Charles Dawson was the lone culprit. Many were outspokenly suspicious of Teilhard as a coconspirator, "not so much on the basis of hard evidence . . . but from an intuitive feeling about this man whom they knew well, loved and respected, but who seemed to hide passion, mystery, and good-humor behind a garb of piety."

Moved by an increasing burden of professional intuition, in 1978 Gould went carefully over the entire Piltdown record. Very soon he found that "nothing excluded Teilhard," but he was also forced to admit the opposite, that nothing beyond Teilhard's presence at the pit "particularly implicated him either." He was losing interest in the priest as a suspect when an obliging fate placed in his hands an unpublished letter written by Teilhard soon after the announcement of the Piltdown exposure. (It was supplied by some friends of Teilhard at an anthropological conference in France.) Dated November 28, 1953, the letter was a response to a request from Kenneth Oakley for information on the digging at Piltdown. No sooner had Gould read it than he saw that it held an inconsistency, "a slip on Teilhard's part," that strongly supported the idea of the priest's complicity.

Shortly afterward, Gould paid a visit to Oxford, where he met with Oakley himself, and the two discussed Teilhard's letter and Piltdown "for the better part of a day." Oakley, it developed, had long ago spotted the same inconsistency but had given it an innocent interpretation. Now, under pressure from Gould, he reviewed the matter and at last agreed that the French cleric might well be implicated. "I think it's right that Teilhard was in it," Gould quotes him as conceding.

The "slip" that Gould picked up in the letter occurs in a short passage where Teilhard offers some comment on the second Piltdown discovery, the one made by Dawson at Sheffield Park (note, however, that the name itself is not mentioned):

As far as the fragments of Piltdown Locality 2 are concerned, it must be observed that Dawson never tried to emphasize them particularly, although (if I am correct) these specimens were announced *after* the finds in Locality 1 were complete. He just brought me to the site of

Locality 2 and explained [to] me that he had found the isolated molar and the small bits of skull in the heaps of rubble and pebbles raked at the surface of the field.

Now this, stresses Gould, quite definitely "cannot be" what happened, not as it is described by Teilhard in those few casual words. Dawson's "discoveries" at Piltdown Two were made during the first half of 1915, but since December 1914 Teilhard had been in the French army serving as a stretcher-bearer at the front. He never again saw Dawson alive, so could not have been taken by him to the new site. For Gould, this glaring contradiction meant only one thing: that Teilhard "could not have seen the remains of Piltdown 2 with Dawson unless they had manufactured them together before he left." Another letter of Teilhard's, in the museum's files, only added to that conclusion.

Though accepting the slip in the November letter as innocuous, Oakley had promptly written Teilhard for a clarification, which was sent in January 1954. As Gould read the explanation in London in 1979, it revealed a mind curiously disconcerted, rather obviously beginning "to temporize" in an effort to recover its balance. Wrote Teilhard:

Concerning the point of "history" you ask me, my "souvenirs" are a little vague. Yet by elimination (and since Dawson died during the first war, if I am correct) my visit with Dawson to the second site (where two small fragments of skull and the isolated molar were supposedly found in the rubbish) must have been late July 1913. I cannot remember whether Smith-Woodward was with Dawson and me, this particular day. But the possibility is not excluded . . .

When I visited the site No. 2 (in 1913?) the two small fragments of skull and the tooth *had* already been found, I believe. But your very question makes me doubtful . . . Yes, I think definitely they *had* been already found, and that is the reason why Dawson pointed to me the little heap of raked pebbles as the place of the discovery.

Before reaching a decision, Gould earnestly sought a way out for Teilhard. If the visit really did occur in 1913, he considered, perhaps Dawson had taken his friend to the second site *before* the finds were made. But Teilhard's second letter blocked that interpretation, saying that Dawson took him there precisely to inspect the place of the discovery, even pointing out the actual heap of rubble in the field. But might Dawson have made the finds in 1913 rather than 1915, Gould asked himself, then for some reason withheld the news for almost two years from Woodward and the others? That made even less sense, since Dawson could not be sure that Teilhard would keep the secret, if secret it was meant to be.

Could Teilhard simply have gone astray in his personal memory of the event, confusing what he later *read* about Piltdown Two with some real but unrelated incident? That seemed most unlikely, especially in view of the graphic detail given in the second letter. No real alternative, Gould at last decided, offered itself for the disturbing theory that the priest and the solicitor together had "planned the Piltdown 2 discovery before Teilhard left England." Before long, there yielded to Gould's eager search a good deal of supporting evidence as to Teilhard's guilt, in his view leaving little doubt of the priest's complicity.

In his November 1953 letter to Oakley, Teilhard actually misstates by some two years the true length of his acquaintance with Dawson. Where in reality they had met first in the summer of 1909, Teilhard writes that he first spoke with Dawson during "a chance meeting in a stone quarry near Hastings in 1911." This error, as Gould sensibly conceded, might be due to a lapse of memory. But he also pointed out how it tended to deflect suspicion from the priest, a feeling that Gould found to be reinforced by another Teilhard letter to Oakley in which he states that "in 1908 I did not know Dawson." Though literally true, this struck Gould as disingenuous since it was only "a few months later" that Teilhard and Dawson met. That fact, he thought, would have been plainly stated by anyone with a clear conscience.

Equally open to question, in Gould's eyes, was Teilhard's attempt in the same letter to portray himself in youth as rather naive and on the pious side, as well as severely restricted in his movements. "You know," wrote Teilhard, "at the time I was a young student in theology, not allowed to leave much his cell at Ore Place (Hastings), and I didn't know anything about anthropology (or even pre-history)." Against this bland assertion, Gould effectively arrayed the many letters to his parents that Teilhard had written while at Hastings. All these letters, quite chatty in nature, he said, reflected a sophisticated outlook on the part of the young priest-scientist, much more than he now admitted. Also, despite seminary regulations, it was obvious that he had enjoyed much personal freedom:

These letters speak little of theology, but they are filled with charming and detailed accounts of Teilhard's frequent wanderings all over southern England. Eleven letters refer to excursions with Dawson, and no other naturalist is mentioned so frequently. If he spent much time at Ore Place, he didn't choose to write about it. On August 10, 1913, for example, he exclaims, "I have traveled up and down the coast, to the left and right of Hastings; thanks to the cheap trains so common at this time of year it is easy to go far with minimal expense."

In all this, said Gould, a definite "pattern" of evasion is glaringly revealed. Further, considerable force and substance was added to the pattern by still another fact, discovered by Gould only after much labor in combing through Teilhard's multivolume *Collected Works*.

That Teilhard in 1920 had contributed a short article on Piltdown to a French science journal was known. What no one realized until Gould pointed it out was the nearly complete silence regarding Piltdown that Teilhard maintained thereafter. Through his final three decades, even as he became increasingly active in paleontology on a world scale, it was as if Piltdown did not exist. After the appearance of the 1920 article, Gould writes,

> Piltdown never again received as much as a full sentence in all his published work (except once in a footnote). Teilhard mentioned Piltdown only when he could scarcely avoid it—in comprehensive review articles that discuss outstanding human fossils. I can find fewer than half a dozen references in the twenty-three volumes of his complete works. In each case Piltdown appears either as an item listed without comment in a footnote, or as a point (also without comment) on a drawing of the human evolutionary tree, or as a partial phrase within a sentence about Neanderthal man.

In itself, and aside from all analysis, thought Gould, this silence of Teilhard's regarding the great event of his scientific youth—an event about which other scientists had written voluminously—was nothing less than "extraordinary . . . inexplicable to the point of perversity." More importantly, Piltdown would actually have provided Teilhard with strong support for his mystico-philosophical theories on human evolution, based on the gradual dominance of spirit over matter, by attesting to multiple lines of hominid development. Yet for more than thirty years he had deliberately ignored and rejected this valuable prop.

Then there was Teilhard's strange reluctance to talk about the hoax during his 1954 visit to London, silence that, in another way, Gould also judged to be incriminating. That year the Natural History Museum had bravely mounted an exhibition explaining the fraud to a fascinated public, and demonstrating how it had been uncovered. All during his stay in London, Teilhard painfully turned aside inquiries on Piltdown, and when taken to see the exhibit at the museum, as Oakley recalled, he "glumly walked through as fast as he could, eyes averted, saying nothing." Before the tour ended, Teilhard's assistant took Oakley aside and told him that "with Father Teilhard, Piltdown was a sensitive subject." Also, as Gould duly noted, Teilhard was the only paleontologist of standing who failed to be stirred by or to respond generously to news of the fraud's exposure. Not until Oakley troubled to write him in New York did he so much as acknowledge his connection with the old event, by then receiving international attention.

For Gould the case against Teilhard became almost airtight when he added the fact of the priest's three-year residence in Cairo (1905–1908) as a teacher of chemistry and physics at a Jesuit boys' school. His presence in Ichkeul in Tunisia (probable source of the radioactive elephant tooth) or Malta (source of the distinctive hippo tooth) could not be proved, though both were possible. But an actual visit to either place was not required. Teilhard's letters from Cairo, noted Gould, "abound in tales of swapping and exchange" with other naturalists in North Africa. The two rare animal teeth, Gould felt, might easily have been obtained from friends familiar with those sites.

To this stage, Gould's handling of the evidence had been adroitly done. In announcing his solution of the mystery, however, he veered rather wildly. Stringing together his clues, he judged the French cleric to have definitely been a guilty party in the Piltdown deception, "an active coconspirator with Dawson." But conspiracies, he added unexpectedly and, as it seems, gratuitously, "have a tendency to spread." He would not, therefore, limit the involvement at Piltdown to just these two: "Once we admit Teilhard into the plot, should we not wonder about others as well?" Hinting that the identity of these "others" might not be beyond reach, he then went a step farther. Strong suspicions had long been entertained, he said, by various unnamed "knowledgeable people" against certain "young subordinates" employed at the time by the British Museum. While none of these are named, most readers would have realized that the list could not be a long one. Wisely, if disappointingly, Gould does not try to unravel the mechanism of this extended plotting, or its purpose, or to say how it might relate to the succession of finds.

In his treatment of the always quicksilverish question of motive, Gould further lost his footing. From start to finish, he suggests, the Piltdown hoax was one enormous joke, harmless to begin with, but which in the end had swerved out of control. Assuming a life of its own, it had finally made the guilty Teilhard himself a prisoner of circumstance, his lips necessarily sealed. As envisioned by Gould, the joke-gone-awry theory depends for coherence on a nice balance of factors, so in fairness his own lengthy description must be given.

Teilhard, he begins, in reality was not the withdrawn ascetic, the aloof mystic, that is suggested by his writings:

> He was a passionate man—a genuine hero in war, a true adventurer in the field, a man who loved life and people, who strove to experience the world in all its pleasures and pains. I assume that Piltdown was merely a delicious joke for him—at first. At Hastings he was an amateur natural historian, with no expectation of a professional career in palaeontology. He probably shared the attitude toward professionals so common among his colleagues—there but for the vagaries of life go I. Why do they have the fame, the reputation and the cash? Why do they sit at their desks and reap rewards while we, with deeper knowledge born of experience, amuse ourselves? Why not play a joke to see how far a gullible professional could be taken? And what a wonderful joke for a Frenchman, for England at the time boasted no human fossils at all, while France, with Neanderthal and Cro-Magnon, stood proudly as the queen of anthropology.
>
> What an irresistible idea—to salt English soil with this preposterous combination of human skull and an ape's jaw and see what the pros could make of it!
>
> But the joke quickly went sour. Smith Woodward tumbled too fast and too far. Teilhard was posted to Paris to become, after all, a *professional* palaeontologist. The war broke out, and Teilhard had to leave just as the last act to quell scepticism, Piltdown 2, approached the stage. Then Dawson died in 1916, the war dragged on to 1918, and professional English palaeontology fell further and further into the quagmire of acceptance.

What could Teilhard say by the war's end? Dawson could not corroborate his story. The jobs and careers of other conspirators may have been on the line. Any admission on Teilhard's part would surely have wrecked irrevocably the professional career he had desired so greatly, dared so little to hope for, and at whose threshhold he now stood with so much promise . . . I cannot view his participation as more than an intended joke that unexpectedly turned to a galling bitterness almost beyond belief.

The notion of such an elaborate, well-planned, and long-continued fraud being nothing more than a joke must trouble any sensitive mind aware of all the facts. The further suggestion that the hoaxer would be unable, for whatever reason, to reveal what he had done seems even lamer. But Gould's tiptoeing among possibilities and half-truths almost manages for a moment to make it seem plausible. For the rest of his life, Gould concludes, Teilhard must have suffered grievously for what he did at Piltdown. The sensitive Frenchman's anguish, he thought, would have been deep and lasting as he watched so many learned and dedicated men, particularly Arthur Woodward and his own mentor at the Paris museum, Boule, strive to explain and defend something that in reality was a colossal fraud.

Surprisingly, the one piece of evidence that Gould puts forth as the solid centerpiece of his argument against Teilhard proves decidedly shaky under attack, not at all difficult to overturn. Moreover, it seems fair to say that once the centerpiece has been eliminated or neutralized, there remains little of a serious nature to sustain a charge of complicity of any sort.

Teilhard's claim that Dawson in 1913 had taken him to inspect the site of Piltdown Two—in reality not discovered until 1915—Gould offers as the priest's "fatal error." Here, he declares, is one of the two "major points" on which his case rested, one of its "two foundations," or as he insists in summing up, "my single strongest argument." But the supposed inconsistency in Teilhard's statement, it turns out, points to nothing more than a confused memory feeding on something that had never been really understood in the first place. The "slip" caught by Gould was no slip at all, not one meant to deceive, a fact quite sufficiently made clear by the existence of the Barcombe Mills materials.

It was to the Barcombe Mills locality, discovered in June 1913, that Dawson took Teilhard, not to Sheffield Park. Writing some forty years after the fact, and while information about both sites was still heavily clouded, Teilhard had understandably mixed up the two places, a confusion that certainly had begun several decades before. Brought to light only in 1949, Barcombe Mills had received little attention in the short interval prior to exposure of the hoax in 1953. The Sheffield Park site itself (not then so named in the public record) was known only from Woodward's original, unspecific reference to it as "one large field, about two miles from the Piltdown pit." In other words, as with everyone else, Teilhard never really knew or understood about the different sites apart from Piltdown itself.

Dawson's report of the first skull fragment from Barcombe Mills was made to Woodward in early 1913. At that time Teilhard was in France, having returned there the previous July. But at the start of August 1913 his superiors posted him back to England—he arrived at Dover on the fourth—assigning him once more to the Jesuit house at Hastings. Stopping off in Lewes on his way, for three days and two nights he was the Dawsons' houseguest at Castle Lodge, and most of that same three-day weekend he spent helping with the excavations at Barkham Manor. At Ore Place on August 15 he began a compulsory spiritual retreat (the main reason for his return to England), which lasted until the twenty-fourth, the continuous exercises keeping him from the excavations for two successive weekends. He was back at the pit on Saturday, August 30, when he found the canine. Within a day or two he once more left Hastings for France, stopping in London to visit the Woodwards and to do some work at the museum. By September 5 he was aboard a steamer crossing the Channel.

At Hastings Teilhard had spent a total of twenty-five days, and for perhaps half that time he was available to accompany Dawson on an inspection trip to Barcombe Mills, which was scarcely a five-minute journey north from Lewes by train. Life at the seminary was indeed restrictive, as Teilhard's partisans insist, but not to the exclusion of frequent outings for fresh air and exercise, especially on weekends after ordination (which for Teilhard had occurred in August 1911).

It was in 1917 that Woodward announced the discovery of a second set of bones at a site which he left unidentified. Though Teilhard was at the front as a stretcher-bearer, it is certain he would have heard this important news, if in no other way than by means of Boule at the Paris museum. That he should have proceeded to connect Woodward's vague, generalized description of the site of Piltdown Two with the field to which in 1913 he had been led at Barcombe Mills is likely and not at all surprising. It becomes virtually certain, moreover, in light of one added circumstance: the fact that the fossil finds at the two sites were nearly identical.

In his 1954 letter to Oakley, Teilhard lists the specimens from what he calls "locality 2" as being an "isolated molar" and "several pieces of the skull." This adequately describes the yield from Sheffield Park, the actual Piltdown Two. But it also coincides with the specimens later listed as from Barcombe Mills (setting aside the small cheekbones): two pieces of cranium and a single lower molar. This close identity of fossils would have further confused the imprecision that was already present in Teilhard's mind about the two sites (and in everyone's mind), and which his memory in 1953 was unable to resolve.

But this curious coincidence in fossil content from site to site must not be passed over too rapidly, for it holds an even deeper significance than that of exonerating Teilhard. From two entirely different sites came almost identical specimens. One set was used belatedly, and the other set was never used at all. Its existence was brought to the attention of science only much later and by happenstance—reason enough for pause! Some as yet unguessed connection, some dim purpose, seems suggested as linking the two sites. What that link may have been, exactly, what the purpose, are questions to be reserved for their proper place in the unmasking farther along. Here the solution may be anticipated to this extent: Barcombe Mills provides the final necessary clue by which the whole elaborate Piltdown design may be traced to its origins. It hints, as well, at just how narrow an escape the young priest had from an even more serious involvement in the developing fraud. If nothing else, in calling forth that particular clue the charge made against Teilhard by Gould has performed a worthy service.

The defenders of Teilhard's innocence tend to plunge headlong into an interminable analysis of elements in the case which, when judged as substantive issues, are seen as largely superfluous. Lengthily and in minute detail they pursue such questions as whether Teilhard, under the strict regimen of the seminary, would have had either the time or the privacy to prepare the fakes, whether he could have slipped away often enough for the necessary visits to the pit, whether and how he could have procured the animal and human remains. But in simple truth, the only serviceable response to such matters is a shrug. *Any* very determined hoaxer certainly could have overcome all these barriers, and others of like nature, and no amount of argument will prove the contrary. Two of these tangential points, however, though they perhaps fall short of conviction, deserve brief notice.

The first concerns a subtle error made by Gould in which he states that Teilhard "could not have seen the remains of Piltdown 2 with Dawson, unless they had manufactured them together before he left." As was quickly shown, Teilhard in his letter to Oakley says nothing about *seeing* the faked specimens in company with Dawson, only that Dawson had taken him to the field of their discovery. His mention in the letter of the molar and the bits of skull, the veritable Piltdown Two remains, could have come from his reading or from visits to the museum on subsequent trips to London, beginning in 1920.

The second correction refers to a false impression fostered by Gould in another of his comments, that in Teilhard's long correspondence with his family "eleven letters refer to excursions with Dawson." This supposed series of "excursions," countered the defense, unfairly implies what might be taken as an "important working contact" between the two men. True, there are eleven letters that mention Dawson, but only four refer to field trips, and three of these to Piltdown. Two others, perhaps three, refer to visits made by Dawson to the seminary at Hastings to inspect Teilhard's collection of fossil ferns, a collection considered important enough for description at a meeting of the Geological Society in London. Most of the collection was donated eventually to the Natural History Museum.

Ephemeral, no doubt, and also to a degree irrelevant, are questions of morality and character in reaching judgment in cases of fraud (aside from courts of law). All the men charged in the Piltdown affair were possessed of sterling public reputations and high moral character, qualities readily invoked by their defenders as proof that the charges could not be true. Conan Doyle's daughter stated flatly that her father's "moral nature" alone, admittedly a lofty one, would rule out his participation in anything low or malicious. But such a defense does not and should not weigh heavily in the present instance, especially not in the face of material evidence to the contrary. Many a good man before this, notwithstanding an exemplary character, has fallen from grace. Yet it can also be argued that there certainly exist people with whom the question of personal integrity and the constraints of a felt morality naturally assume a more than ordinary role. If so, a figure such as Teilhard de Chardin surely is one.

His entire life was spent as an ordained priest, always at least outwardly obedient to the commands of his church. The intellectual opinions that brought him fame concerning man's origin and destiny on earth, difficult and controversial as they are, even now hold some part of the world's attention. It is something to remember that those challenging opinions were already in full ferment in the days of Piltdown, agitating the mind of the young Teilhard as he hunted fossils along the Hastings cliffs and at Barkham Manor. Whether his theories on human evolution will eventually be vindicated as seminal and true, or set aside as excessive and mistaken, is not relevant here. The task at the moment is to inquire whether a man of declared high principle whose whole aim in life was the pursuit of intellectual and spiritual truth, could have stooped to massive deception—and this in the very discipline that occupied his most ardent hopes and thoughts.

Is it credible that the man who penned the following autobiographical passage was, at the very same moment as he here describes, plotting and carrying out the Piltdown forgery:

> It was during the years when I was studying theology at Hastings (that is to say, immediately after I experienced such sense of wonder in Egypt) that there gradually grew in me, as a *presence* much more than an abstract notion, the consciousness of a deep-running, ontological, total Current which embraced the whole universe in which I moved; and this consciousness continued to grow until it filled the whole horizon of my inner being . . .
>
> At first, naturally enough, I was far from understanding and appreciating the importance of the change I was undergoing. All that I can remember from those days (apart from the magic word 'evolution' which haunted my thoughts like a tune: which was to me like an unsatisfied hunger, like a promise held out to me, like a summons to be answered)—all that I can remember is the extraordinary solidity and intensity I found then in the English countryside, particularly at sunset, when the Sussex woods were charged with all that 'fossil' Life which I was then hunting for, from cliff to quarry, in the Wealden clay.

There were moments indeed when it seemed to me that a sort of universal being was about to take shape suddenly in Nature before my very eyes . . .

Because those fervent words were set down toward the end of Teilhard's life, it may be thought that they do not truly mirror his state of mind at Hastings some forty years before. But an essay he wrote during World War One, in periods of quiet at the front, shows that even then, and for years previously, he had been accustomed to indulging in such world-encompassing ruminations. The essay, "Cosmic Life," was completed in April 1916, when he was thirty-five, some three months before the death of Dawson. It clearly and strongly foreshadows much of his later philosophy.

There remains the second of Gould's two major points, the curious fact of Teilhard's thirty-year silence about Piltdown in his writings. (To be clear on this, after his 1920 article in an obscure French periodical, Teilhard's lengthiest reference, by far, to Piltdown is a small paragraph of less than two hundred words—a dozen lines—occurring as a footnote to his discussion of the Heidelberg jaw in his 1943 paper, "Fossil Men." It questions the linkage of the Piltdown cranium and jaw.) Sincere, even strenuous attempts have been made by Teilhard's adherents to explain this wholesale reticence, all of them tied to technical arguments over the priest's views of the fossil record and its interpretation. Without exception, however, these arguments fall short of meeting the problem posed by such unrelenting neglect of a topic that, in any reasonable view, should have been prominent.

A few commentators, on the other hand, have squarely faced what they see as the true implication of this silence, the intriguing possibility that Teilhard suspected, definitely knew, felt, or guessed that something was wrong at Piltdown and that actual fraud was somehow involved. As one critic wrote, for Teilhard even to admit such an eventuality must have left him feeling "deceived and cheated. He loathed, no doubt, to suspect anybody, and did not want to have anything to do with the case." Certainly a burden of worry over the authenticity of Piltdown, vague as the worry may have been, would explain the glum distaste Teilhard made no effort to hide from Oakley and the others while he was in London in 1954.

In fact, there does exist some evidence, a little more than circumstantial, to show that Teilhard may indeed have entertained fears that somewhere in the Piltdown record there lurked an element of fraud. In his 1920 article he expressly joined with those scientists who still refused to admit that there could be any connection between the jaw and the cranium. For him, as he said, both pieces were legitimate fossils, but the jaw was that of an animal, without connection to the other bones. Only the cranium had relevance for human descent, only the cranium was the true *Eoanthropus dawsoni*. Up to his death, though never writing about it, on the basis of the cranium Teilhard retained Piltdown in his schematic drawing of man's evolutionary tree, but always inconspicuously and without comment. It was out of this quiescent situation that Teilhard, thinking along lines very similar to those that alerted Weiner, perhaps formed his first nagging suspicions.

Though jaw and cranium were *found* together, they did not *belong* together. The two rare items coming together in the ground was a downright unlikely occurrence, even radically improbable. Add to this the fact that the jaw's two most crucial parts, the condyle and the symphysis, were both missing. Then add the belief that apes had never inhabited Pleistocene England, and discard as ridiculous the idea that a passerby had dropped an unwanted fossil ape jaw into the pit. At some point, inevitably, thoughts like these must strike a spark about possible human agency or fraud (especially remembering that fossil fraud was not unknown, beginning with the famous Moulin Quignon deception fifty years before). A trigger for the notion would hasten matters, and in Teilhard's case there was indeed such a trigger: as can be shown from his own words, for many years he carried in his memory one tiny seed of actual suspicion.

In his initial letter to Oakley of November 1953, Teilhard comments on the idea of deception at Piltdown. In the course of his statement he offers no less than three references, more or less overt, to Dawson's possible complicity. While the first two in a casual reading are perhaps easily overlooked, the third when studied in context becomes almost flagrant:

Of course nobody will even think of suspecting Sir Arthur Smith-Woodward. But to a lesser degree this holds for Dawson too. I knew pretty well Dawson, since I worked with him and Sir Arthur three or four times (after a chance meeting in a stone quarry near Hastings in 1911). He was a methodical and enthusiastic character, entirely different from, for instance, the shrewd Fradin of Glozel. And, in addition, his deep friendship for Sir Arthur makes it almost unthinkable that he should have systematically deceived his associates for several years.

When we were in the field I never noticed anything suspicious in his behaviour. The only thing that puzzled me, one day, was when I saw him picking up two large fragments of the skull out of a sort of rubble in a corner of the pit (these fragments had probably been rejected by the workmen the year before). I was not in Piltdown when the jaw was found . . .

The incident that "puzzled" him is reflected in a second letter, written in December 1953, some ten days after the first, to his friend Abbé Henri Breuil. Much shorter, it stresses the fact that the two pieces came from disturbed soil: "The big fragments of skull I did see Dawson find were *not in situ*, but thrown aside into one corner of the pit." Teilhard doesn't mention it, but planting a fake in worked gravel that had been "thrown aside" would of course be easier than putting it into apparently unworked soil.

Woodward's innocence, says Teilhard in the first letter, is unassailable. But that of Dawson, he adds for no stated reason, is less so: in other words, Teilhard *can* conceive of Dawson as author of the fraud. The mention of Glozel, a famous and more recent fossil hoax in France (in 1924), is arresting. Comparing Dawson with Emile Fradin, one of the Glozel perpetrators, Teilhard depicts his friend as of an open if painstaking temperament, where young Fradin was supposedly devious and calculating. But in this context the comparison has no real point or purpose, serving for little else than to bracket the Englishman with the accused French hoaxer. At this juncture, having raised at least a small doubt about Dawson personally, Teilhard then unexpectedly throws open a window on a quite specific detail, greatly increasing the doubt.

While at the pit with Dawson, Teilhard writes, there was never anything "suspicious" about the other man's activities. But then he proceeds to set down a fact—something he says he *saw*—that blatantly contradicts what he has just said: "The only thing that puzzled me, one day, was when I saw him picking up two large fragments of the skull out of a sort of rubble in a corner of the pit." In all Piltdown literature there are few passages to match this one for tantalizing peculiarity. (Provocative enough in itself is the subtle interplay between the words "suspicious" and "puzzled.")

Why should Dawson's finding skull fragments in a spoil heap ("rubble in a corner") be seen as puzzling? That is exactly what happened any number of times during the excavations as each of half a dozen new cranial pieces (five of them spotted by Dawson) came to light. Evidently Teilhard's statement is incomplete. Missing is some fact which, when added, will make the sentence comprehensible, and what immediately comes to mind is this: on no single day during the entire excavation did Dawson or anyone else report finding *two* skull fragments. Yet that is what Teilhard claims he saw, a claim repeated in the letter to Breuil.

As it happens, it is still possible to calculate about how often Teilhard was at the pit, and on which days. In 1912 he was definitely on hand for most of Saturday, June 2 (he departed early), then spent the next two weekends on assignment at Bramber. On the weekends of June 30 and July 7 he was at Hastings, and free, so he could have gone to the pit, though there is no indication that he did. By mid-July he was back in France. In 1913 he helped at the pit on only four days, the long weekend of August 8–10 and on Saturday, August 30. During all those days, nine at the most, the record shows that only once was any part of the skull found by Dawson, and that was on June 2, when he picked up one small piece of the occiput (the same day that Teilhard found his *Stegodon* tooth, which so delighted Woodward). But Teilhard plainly says that on some unspecified day he saw Dawson take from the rubble no less than two fragments, and that both were "large."

The question may be repeated: what was it that *puzzled* the young priest? Obviously it was something far enough out of the ordinary to have remained alive in a corner of his mind for forty years. The solution to this miniature mystery requires only a slight shift in the angle of vision.

Teilhard says that as he looked, Dawson picked up two fragments of the skull. More probably the watching Teilhard caught a glimpse of Dawson as he attempted to *plant* the two pieces of bone in the rubble, an attempt that failed (did Dawson perhaps sense Teilhard's quizzical glance in his direction?). Thereafter the puzzled Teilhard—becoming aware that at no time had more than a single piece of skull been uncovered—was never quite sure of what he saw that day. Why, he must often have wondered, had nothing ever been heard of those two fragments?

As a result, his later hesitation would have centered on just that irritating indecision of never being sure *what* Dawson was doing as he bent over the rubble heap. Further obscuring the fleeting incident would have been the fact that it happened, as Teilhard notes, "in the pit," rather than above ground, where the rubble ordinarily was deposited.

In his letter of 1953 to Oakley—the conclusion really presents itself as unavoidable—Teilhard purposely included that bare but provocative mention of Dawson's odd behavior, and with one intention only: he hoped that Oakley or someone else would be moved to pursue the matter. For whatever reason, apparently no one did.

In this same connection, Gould contributes a further intriguing possibility about an idea that came to him, he says, in a leap one night as he sat reading Teilhard's 1920 article. Discussing the jaw-cranium linkage, the article makes the statement, so frequently made by others, that if only the absent condyle had been preserved, the pivotal question of jaw and cranium belonging together would have been set to rest long ago. Gould goes on:

> I read this statement in a drowsy state one morning at two o'clock, but the next line—set off by Teilhard as a paragraph in itself terminated by an exclamation point—destroyed any immediate thoughts of sleep: "As if on purpose the condyle is missing!"
>
> *"Comme par exprès."* I couldn't get those words out of my mind for two days. Yes, it could be a literary line, a permissible metaphor for emphasis. But I think that Teilhard was trying to tell us something he didn't dare reveal directly.

As if on purpose. In that brief but evocative phrase, Teilhard's advocates see nothing but the expression of an innocent chagrin at the frustrations imposed by fate. Yet the words do seem to imply more than that. When paired with the similarly teasing phrase in Teilhard's letter—"the only thing that puzzled me"—they grow rapidly in significance. Taken together, the two phrases certainly document something like a deep indecision on Teilhard's part. From that it is no great leap to seeing behind or beneath the indecision a long-standing if unfocused suspicion about the man who befriended him in that Hastings quarry.

It is even possible to suggest—no more than that—where and in what circumstances Teilhard's first faint apprehensions about a fraud may have dawned on him.

After the second season of digging, Teilhard left England for France in September 1913. On the way he stopped for several days at the London home of Arthur Woodward. There, as he wrote soon afterward, he "spent an enjoyable evening *tete a tete* with Woodward, a certain Gregory (from the New York Museum, an important contact), and an ornithologist from the British Museum, Pycraft." The Gregory he mentions as having a part in that pleasant gathering was William K. Gregory, of the American Museum of Natural History in New York. Later that same fall, Gregory's article on Piltdown (quoted above) appeared in the American Museum's own journal with its candid mention of those London rumors about Piltdown being "a deliberate hoax." In talking that evening with the other three scientists, Gregory must surely have alluded to those floating doubts about the Piltdown specimens having been "artificially fossilized," as he expressed it. Even if he raised the disturbing topic only to dismiss it, as he does in the article, his words about a possible hoax must have fallen with a depressing weight of foreboding on Teilhard's listening ears.

Gould's hypothesis makes Teilhard an accomplice of Dawson. Without any attempt to describe their respective roles, or to show how their motives might have dovetailed (no easy task), he made the two men co-conspirators. Then for good measure he widened the plot to include other conjectured but unnamed parties as accessories. As has been demonstrated, there is literally no evidence to connect the priest with Piltdown as a forger in any capacity. Still, the cry of "conspiracy!" is always a difficult one to confute, precisely because it is so shapeless and slippery. Even in the face of proved innocence there will no doubt be observers who insist that Teilhard in some dark moment may well have decided to lend Dawson a helping hand.

Showing, perhaps, what yet may be expected in that line is some more recent speculation about Teilhard, freewheeling indeed, that appeared in the respected *Antioch Review* for fall 1986. This article goes so far as to make Teilhard the originator of the fraud, with an innocent Dawson at first as dupe, then later joining forces with the scheming cleric. It seems that the Piltdown cranium was an authentic fossil, after all, not the fake it was shown to be by the experts. It seems that the skull really *was* found by a sincere Dawson, with the unscrupulous Teilhard taking over and building on it the entire hoax. His motive? To further his philosophical theories. The article's loose approach to its subject is made sufficiently evident by one fact alone: there is no mention whatever of Barcombe Mills.

More recent yet (1992), and perhaps even more typical of what is to come, is an attempt by a Canadian anthropologist, Norman Clermont, to erect a French conspiracy at Piltdown. Though built upon the veriest gossamer, Clermont's argument still managed to find a home in the authoritative pages of *Current Anthropology*. Citing what appears to be another egregious "slip" in a Teilhard letter, Clermont proceeds to incriminate both the priest and his mentor, Marcellin Boule.

In 1937 Teilhard sent greetings to his old master on the occasion of Boule's scientific jubilee. In one passage of the letter, charges Clermont, Teilhard betrays the fact that "he knew about the Piltdown canine 13 months before he was to find it." As translated from the French by Clermont the passage reads:

> Do you remember our first meeting about mid-July 1912? . . . On that day I came in timidly, about two o'clock in the afternoon, and rang at the door so often passed through since of the laboratory in the Place Valhubert. You were on the eve (sacred!) of your departure for holidays, and you were very busy. Nevertheless, Thévenin disobeyed the order. You let me in for all that. And then, with the help of the canine from Piltdown (and also, to be sure, because I too am from the Auvergne), you proposed that I should come and work with you, in the Gaudry, in your own school . . .

Clermont does not know quite what to make of this strange and, it would seem, pointless anticipation of the canine's discovery (found in August 1913), unless Boule himself were taken to be part of the scheme. His brief discussion notes that Boule visited England in the fall of 1912, and he asks portentously, "Was there after all a French connection?" But here again, as so often, the true explanation is utterly simple and devoid of harmful implication.

Clermont's reading of Teilhard's letter is both hasty and insensitive, failing to allow for the passage of time obviously encompassed by the transitional phrase "And then, . . ." Unthinkingly he takes what follows that phrase to be a part, a continuation, of the 1912 visit to Boule mentioned before the phrase. In reality, of course, it refers to the events of 1913, *after* Teilhard's finding of the canine. In the flow of his letter (written twenty-five years after the fact), the priest has merely elided time, unconsciously dropping a phrase, as might any letter writer in the familiar or conversational mode. Properly he should have written, "And then, about a year later . . ." In any case, Boule did not need to be reminded of the true sequence of events, nor was a recital of incidental facts the purpose of Teilhard's letter.

But innocence or guilt will never be established through fine-drawn analyses of matters that are, to start with, so convoluted and questionable. As with Conan Doyle, it is in exposing the true forger and revealing his methods—especially concerning the many planting operations at the pit—that Teilhard is fully vindicated.

❧ 10 ❧

The Anatomist Accused

A few weeks before his death in January 1955, Sir Arthur Keith wrote to an old friend about Piltdown. Almost a year had passed since public revelation of the fraud and the visit paid him by Weiner and Oakley. Repeatedly during that year, newspapers had asked him for his comments on the hoax, but he had consistently refused to speak. "In a complex business like this," he said warily if not cryptically, "the only thing is to be tongue-tied." Then, a week after his passing the letter to his friend was quoted from at length on the front page of the London *Times*. In it, as the paper pointed out, Keith left no doubt about where he felt the blame for the hoax should lie.

Describing the visit of Weiner and Oakley, Keith recalled how he had readily welcomed them, agreeing that "they had given a very complete solution of the Piltdown problem, although not a happy one, for it left me in no doubt that the man I had the greatest reverence for had deliberately misled his best friend, Smith Woodward, and me." Originally, Keith had accepted the jaw as authentic "on the strength of Dawson's deliberate word" that it had come from the pit's deepest levels (apparently forgetting for the moment that Woodward had also been present at the jaw's discovery). While it was true that some scientists had rejected the jaw, he went on, "no one suspected fraud on the part of the discoverer. That discovery was not made until 40 years afterwards by one who had never encountered the 'honest' countenance of Charles Dawson."

Keith's conviction that had it not been for Dawson's genial, ingratiating ways he might have succeeded in avoiding the embarrassing trap was a nagging thought he expressed over and over in the last months leading to his death, seemingly not caring how the admission reflected on his abilities as a scientist. "Never was a man so plainly stamped with honesty in word, in face, in manner; such a simple-hearted, honest fellow," is a passage from another letter to a different correspondent in August 1954. Several weeks after that, writing to a third correspondent, he expresses the same thought in even stronger terms: "So compelling was the honesty of Dawson's manner of speech that not a single soul of us doubted his word. Those who solved the Piltdown problem are of a generation who never fell under the spell cast on all by the living Dawson . . . I have no doubt that he was the author of all the fraud." What would have made the taste of disillusionment even more bitter for Keith was the memory of how much glowing praise he had heaped on Dawson over the years—in letters, by word of mouth, in an obituary, in his varied writings on Piltdown, at the 1938 monument dedication, in his foreword to Woodward's book, and in his own 1950 autobiography.

Arthur Keith went to his grave believing, or at least proclaiming, that in large part he had been the innocent victim of a malevolent dissembler. During the next thirty-five years, as the list of possible perpetrators grew steadily, with Dawson at its head, not the least breath of suspicion arose against Keith himself. But at last in 1990 came the anatomist's turn. Far from having been the prey of a disarming personality, it was charged, Keith had actually served as the cunning mastermind of the whole fraudulent enterprise, directing the surreptitious activities of the man he had spent his last days condemning. Part of the evidence lodged against him was the very praise he had lavished on his supposed cohort.

The case against Arthur Keith begins with the work of an Australian science historian, Ian Langham. Upon Langham's early death in 1984, well before he had a chance to publish any of his findings, his work was continued (at the request of Langham's family) by another science historian, Frank Spencer, an Englishman teaching at Long Island University in New York. The book setting forth the Langham-Spencer thesis, written by Spencer and published by Oxford University Press, was an oversize volume offering the most comprehensive and detailed history of the Piltdown fraud and its background yet published. In one lengthy chapter, supported by extensive notes, Keith's role in the story is investigated, and he is judged guilty "beyond a reasonable doubt." The charge received worldwide publicity.

As a whole, Spencer's volume and his presentation of the case are impressive, especially for scope and detail. Yet the commentary on the book in reviews and articles was split rather evenly between agreement and dissent. Then in June 1992 the Keith case was taken to another level of professional interest when the journal *Current Anthropology*, leader in its field, published in its pages a very long article—equal in length to a small book—restating and strengthening the charge. Written by a prominent South African anthropologist, Phillip V. Tobias, the article was accompanied by commentary solicited from a number of other scientists. Here again opinion among the experts was split, and in addition there now showed up much wavering uncertainty.

Between them, however, and despite the steady if controlled fire of dissent, the Spencer book in tandem with the *Current Anthropology* treatment have managed to elevate Keith to the position of principal suspect.

Ian Langham, then of the University of Sydney, long fascinated by the Piltdown mystery, had at first chosen a different man as culprit, Grafton Elliot Smith. In 1978 he produced a long and circumstantial article purportedly proving Smith's guilt. But a year or so later while in London he stopped in at the Royal College of Surgeons, Keith's old employer. There, in the college's archives, Langham unexpectedly came across evidence that, for him at least, pointed in only one direction: not Smith, but Sir Arthur Keith was the Piltdown forger. His subsequent studies, followed by those of Spencer and Tobias, have produced a case which, more than most of the others, exhibits a peculiar tendency toward confusion in chronology and in the handling of details. There is also present an inclination to hurry over some crucial points, where in reality a calm patience and a more deliberate pace are needed. In the following review, therefore, a heightened concern for precision and a more minute analysis will not be out of place.

At his death Sir Arthur Keith left his personal papers to the library of the Royal College of Surgeons. Included was a diary he had kept intermittently before and during the Piltdown period, in which every few weeks he made round-up entries. Two of these entries, one brief, the other longer, relate to Piltdown concerns and are what aroused Langham's suspicions against the diarist.

On Sunday, December 22, 1912, a brief entry sums up the high points of the preceding few days:

> This has been an exciting week. First my paper on the evolution of the mammalian lung at S.K. (2) The famous meeting of the Geological Soc. on 18th: crowded room. I write leader for B.M.J. on the meeting Monday night (16th); on Wednesday wrote acct for Morning Post (get home at 12) dined with Reid Moir. On Thursday long interview with Manchester Guardian: thus keeping things as straight as I could during the week and thrusting a quiet and fairly effective spoke in the Boyd Dawkins and Smith Woodward wheel. I expect it will be war to the death now between the R.C.S. and S.K. people. Now settling down to put papers in order for lectures on upright posture . . .

The article ("leader") Keith wrote for the *British Medical Journal* treated the famous Geological Society meeting held at Burlington House on December 18 to announce the Piltdown discoveries. According to Keith's own notation the article had actually been written two days *before* the meeting, on December 16. It made its appearance anonymously in the *Journal* on Saturday, December 21, under the heading "Discovery of a New Type of Fossil Man." Its ten long paragraphs, set in fairly small type, fill three whole columns. Opening with a brief reference to the Society's "quiet meeting room" as being "crowded" that night, as well as being charged with an "air of expectancy" as Woodward and Dawson took the dais, it gives no specifics of the meeting. Instead the text consists mostly of technical discussion of the Piltdown specimens. Two paragraphs, however, relate to the circumstances of the discovery itself, and it is here that questions arise.

The article's explicit information, declare Keith's accusers, especially as to the nature and location of the pit and some other detailed items, could not have been known to him when he wrote the piece on the sixteenth unless he had a special source of information, or unless he had been to the pit himself. The article's pertinent portion reads:

> Leaving the question of antiquity still undecided, we shall now turn to the narrative of the actual discovery. The scene of this "find" lies some

nine miles north of Lewes, in the valley of the Sussex Ouse, which, rising in the Weald, breaks through the South Downs at Lewes, and enters the sea at New Haven. After flowing eastwards past Sheffield Park the Ouse bends southward. On the north bank, at the bend, about a mile from the river and on a flat field near Piltdown Common, in the parish of Fletching, situated 80 feet above the level of the river, there is a superficial bed of gravel 4 ft. thick. It is in this bed of gravel that the fossil bones were found by Mr. Charles Dawson of Lewes . . .

As an outsider, Keith would have enjoyed no access to privileged sources, yet here he was, two days before the announcement, supplying any number of specific references about the location, layout, and nature of the discovery site. His only other source for the information, it was said, would have been personal knowledge of the Piltdown area gained on some prior inspection trip. But at that point Keith, it was certain, had never visited Barkham Manor. The second diary entry found by Langham here assumed a peculiar importance.

On Sunday, January 5, 1913, Keith entered in his diary a short account in his usual staccato manner of a trip he and his wife had made the day before to Piltdown. The purpose was to view the actual site, the original gravel bed. Yet strangely enough, while the couple came close to it, they seem not to have reached the pit itself:

Yesterday Celia and I got up at 7. Caught the Uckfield train L. Bridge at 9.7 Uckfield in hollow of tributary valley of Sussex Ooze [Ouse]— at 11.20: lunch at Maiden Head [Uckfield]. Head westward on foot across another trib. valley and by one-30 were on the Piltdown Common—heath form of land . . . Through village of Piltdown leaving Fletching on our right, crossed the slow valley of the Ooze and up to the village of Newick on other side. Found circular route could not be carried out owing to the state of paths came back by Piltdown—boys told us where Sussex skull found: fir avenue leading to farm—white gate: on Delta plateau above the Ooze. Didn't see the gravel bed anywhere. Back at the Maidens Head 4:15. Tea and warmth . . . walked upwards of 12 miles and tired.

Having walked so far to seek the pit, asked Spencer, why should Keith have "abruptly terminated his search at the gate . . . why did he not walk the couple of hundred yards to the Manor House and complete his mission?" The answer to that apparently legitimate question, according to the Langham-Spencer theory, supplied a devastating instance of Keith's real cunning: the diary entry itself was spurious. It was intended to establish the lie that Keith's *first* journey to the Piltdown vicinity had been made on that luckless January 4 trip with his wife. Certainly his inability to find the pit, even with the directions given him by the locals, fostered the notion that he had no knowledge of its location.

For Langham and Spencer, the spurious nature of the diary entry was made nearly positive when they went carefully over still another article on Piltdown written by Keith at about this same time.

In his diary Keith customarily—*not* always—made brief notations about the sale and publication of his many writings, books and articles. Several such references to the London monthly *The Sphere* were spotted by Langham, especially one concerning Piltdown—"Our Most Ancient Relation"—which appeared in the September 1913 issue. Looking through this article in the volume for 1913, Langham came across a second article on Piltdown published in *The Sphere* that year, but earlier, in January, some two weeks after Keith's abortive search for Barkham Manor. While it was unsigned, and no diary entry confirmed the supposition, both instinct and internal evidence convinced Langham that Keith was its author. In it an unidentified "correspondent" reports that on a visit to Piltdown "a few days ago" he had the pleasure of gazing down into the famous pit. Note that the correspondent arrived at Uckfield by train, and had then set out for Piltdown on foot, as had Keith:

I had never heard of this Sussex Common until the discovery of the skull, although the heath can boast a golf course. The easiest way to reach it is by main-line train to Lewes, and thence by motor-train to the little hillside town of Uckfield . . . the road to Piltdown pitches up and down due west from the town . . . I kept on the left-hand road until I approached a spot marked "Barkham" on the ordnance map and then struck across the common towards an avenue of firs. At the side of this avenue was a ferruginous-looking cutting or excavation, simple enough in appearance, but indeed the shrine for which I was making. Closer examination of the gravel showed that it was promising-looking ground . . . It was strange as one stood at the side of this little trench to think of the interest this spot has created in scientific circles . . .

If indeed it was Keith who stood at the side of the "little trench" (and very probably it was), just when had this particular visit taken place? Certainly, it could not have been on January 4, a fact made clear by the diary entry for that date. Nor was it, judging by several subsequent diary entries, that had him busy elsewhere, between January 5 and January 18, when the *Sphere* article was published. Therefore, reasoned Langham, Keith's visit to Piltdown, as described in *The Sphere*, must have taken place *before* January 4, 1913. But the diary entry for that date implicitly claims that Keith had never before been to Piltdown—and in fact knew so little about the area that he had gotten lost. Here was what seemed an obvious contradiction, and for Langham at this juncture there was little question "that Keith was in some way involved in the Piltdown affair." This conviction was greatly strengthened when Langham suddenly recalled something said in her old age by the daughter of the Barkham Manor tenant.

Twenty-seven years old at the time of Piltdown, Mabel Kenward lived until 1978. Several years before she died, she gave to Kenneth Oakley an account of a curious but undated incident that had taken place at the pit site. Early one evening while watching from the house, she stated, "I saw this tall man come up, not even up the drive, but across the fields—must have gotten over the hedges and ditches even to get there." The man walked directly to the pit where he "started scratching about." Always alert for intruders on the property, especially at the pit, she came out and accosted the man. "I said, excuse me, are you an authorized searcher?" Surprised at being caught, the man "didn't say a word . . . might have been a ghost . . . and off he went the same way he came across the fields." He was in his forties, she thought, and was wearing "an ordinary grey suit but he had gumboots on and he was very tall." Here, Langham felt sure, was an accidental encounter with Keith as he attempted to pay the site a clandestine visit. The description fitted him perfectly.

Returning to the *British Medical Journal*, Langham was now struck by several other items in the Keith article that had not been known to outsiders before the meeting on the eighteenth and the resulting newspaper reports. There was a reference to the skull as "a thing like a cocoa-nut." There was a precise dating of "four years ago" for the time of the cocoa-nut incident. There was identification of a "rubbish heap" as yielding most of the cranial fragments, as well as the fact that the jaw had come from the pit itself, "the lower part of the stratum." Where could Keith possibly have learned all these things before the general announcement and the resulting public discussion? It was at this point of his investigation that Langham uncovered what seemed another pertinent and highly revealing fact: an apparent attempt by Keith to hide an earlier acquaintance with Charles Dawson.

In his autobiography, published in 1950, Keith names the time of his initial meeting with the solicitor as "One morning early in 1913." Arriving at his office in the College of Surgeons, "I found a gentleman waiting for me. He introduced himself as Charles Dawson. We had a pleasant hour together." In his diary at the time, that same visit is placed more specifically in late January ("Charles Dawson came in to see me at the college—a clever level headed man"). The reason for Dawson's visit to the college that morning, as Keith recalled it to Weiner in 1953, was to explain why, despite Keith's admitted leadership in the field, he had taken the Piltdown specimens to Woodward (their long association seemed to demand it).

Now Langham lit on another entry in Keith's diary, and in pursuing its implications he was able, so he claimed, to establish that Keith and Dawson had actually met no less than a year and a half earlier than Keith was willing to admit.

In a catchall entry dated July 15, 1911, Keith scribbled the sentence, "On Thursday went to Brighton Museum Association." A search of the newspapers for this date showed that it referred to the annual congress of the British Museums Association, that year held at Brighton, on the south coast just west of Hastings. Some further searching, according to Spencer, revealed the fact that the association, as part of its official program,

> ran an excursion to Hastings, hosted by Ruskin Butterfield, Charles Dawson, Lewis Abbott, and other members of the Hastings & St. Leonards Natural History Society and Hastings Museum Committee. According to the *Hastings & St. Leonards Weekly Mail & Times* (15 July 1911) the guests of honor included Francis Bather from Woodward's department at the British Museum (Natural History), Reginald Smith from the British Museum (Bloomsbury branch), and "Mr Arthur Keith, Royal College of Surgeons." . . . Evidently during this visit Dawson and Abbott gave a guided tour of the archaeological sites around Hastings, and later in the day Dawson lectured on the "His-

tory of Hastings Castle." If they had not met before, there could be no
question that they had met that day.

In addition, it could fairly be said that if this undoubted meeting of
the two in 1911 had been covered up, then it became obvious that they
might actually have become acquainted at any time before that.

For Langham, a final piece of the puzzle now fell neatly into place.
It concerned a peculiar cover arrangement by which Dawson was able,
without arousing suspicion, to confer with Keith in his office at the
college.

In the midst of the Piltdown excitement, it appeared, Dawson had
not hesitated to take up a new project—a technical one that might have
seemed well beyond his capacity—involving an anatomical study he referred to as "the 13th Dorsal Vertebrae." In humans the normal number of such vertebrae is twelve. In apes it is thirteen. Among the
exhibits under Keith's supervision at the Royal College were a number
of Eskimo skeletons, some of which possessed the extra vertebra.
Studying these, Dawson in late spring 1912 produced a short discussion
paper comprising eight typed pages and accompanied by photographs.
As he explained to Woodward, he believed that he had initiated "a new
subject," which he hoped would earn the attention of the Royal Society.
Langham, however, found that the subject was by no means new. A
whole book on the topic had been published in France that same year—
and Keith himself had been asked to prepare a notice of it! Keith did
write a review, calling the book a "really valuable work," but not until
much later, 1915, and then his comments on the "valuable work" occupied only ten lines. About Dawson's paper nothing ever happened, and
Langham concluded that the thirteenth vertebra had been nothing but
a "camouflage," suggested and guided by Keith.

The actual working out of the hoax, as envisioned by the Langham-
Spencer thesis, admittedly is rather murky, in some degree even uncertain and incomplete. It has the nefarious Dawson, driven by vaulting
ambition, initiating the fraud entirely on his own, with the unprincipled
Keith being drawn only later into the "extraordinary enterprise." In
Keith's case as well, self-serving ambition is seen as the spur, but it was a
two-fold ambition, touching both his personal and professional life.

Though by 1912 Keith had reached the very top of his profession,
the one remaining distinction he had failed to garner, and that he con-

sidered essential to proper recognition of his work, was election as a fellow to the Royal Society. Twice he had been proposed, and twice had been refused, the reasons, of course, not stated. The public notice that came to him over his integral part in the Piltdown controversy vastly enhanced his professional and public standing—attendance at his weekly lectures at the college tripled—and in the spring of 1913 he at last received his Royal Society fellowship. There was also an added benefit for Keith, for Piltdown Man, with its apparently human brain, lent strong support to Keith's own view of evolution, in which he had predicted a much higher antiquity for modern man than had most of his fellow scientists.

At the start, wrote Spencer, "a suitable skull" was somehow procured by Dawson, "an unusually thick one, perhaps an Australian aboriginal skull, which he then placed in the gravel bed and awaited its discovery." The skull was duly uncovered by the workmen, then promptly shattered by the blow of a pickax, and Dawson spent the next couple of years "recovering the fragments in preparation for their subsequent (?) treatment and replanting." It was at this point, says Spencer, that Keith entered the picture, in fact shortly after his meeting with Dawson in Hastings in the summer of 1911 at the museums congress. But how and why it happened, and exactly what immediate pressures brought England's leading anthropologist to join forces with the wily solicitor in a fraud already begun, goes unspecified. Spencer merely states that when Dawson brought one or more of the skull fragments to Keith, "the discussion naturally turned to the kind of jaw that would be associated with such a thick and apparently very ancient cranium—and it was from this 'crucible' that the forgery emerged in its final form."

That is all that Spencer felt able to venture by way of explanation, allowing conjecture to derive what it could from that vague and inadequate term "crucible." He was content to leave unasked, as well as unanswered, the vital question of which of the two men first broached the idea of cooperation in the fraud, by no means a minor point. Could Dawson, for example, have risked Keith's refusal? But if Dawson did not make the first move, how could Keith have known that fraud was in the air?

Now Keith took control, says Spencer. He prepared all the subsequent Piltdown specimens, including those from Sheffield Park and Barcombe Mills. He further doctored the cranial pieces, and may or may not have begun by "substituting" another skull for the one Dawson planted. The bone implement, the so-called cricket bat, was also his idea, and the jaw was his "master-stroke." The canine, with its coat of ordinary paint, Spencer felt, was produced by Dawson "acting without Keith's approval," perhaps after a quarrel over some point of disagreement. The bones labeled Piltdown Two and Three might have been part of the original plan, but more probably they were produced in a spirit of "crisis management" when questions arose about the linkage of jaw and cranium.

As for Dawson, in Spencer's view he was to be "the man on the spot: responsible for delivering a predetermined schedule of finds and monitoring events within Woodward's inner circle." The complete success of the fraud is a measure of how well he carried out his precarious duties.

Keith's assigned role in the burgeoning fraud, as limned by Spencer, was a more public one, and equally important. He was to "promote the finds and nurture the debate. The greater the debate, the greater the rewards." As events show, this conspicuous part Keith played to the hilt, repeatedly arguing over proper reconstruction of the skull and deliberately picking loud, public fights with Woodward and Elliot Smith. In these purposeful embroilments he sometimes even allowed his performance to verge on the obnoxious. The inevitable result was exactly what he and Dawson desired, ever-increasing attention from press and public at home and abroad.

The old dictum (first formulated by Poe) about unusual features in a mystery making it easier rather than harder to solve, holds true for the case of Sir Arthur Keith. The unusual feature here will be obvious to all: it is that early morning journey of Keith and his wife from London to Piltdown, which supposedly ended in confusion and disappointment.

The circumstance that the anatomist and his wife, after trudging so many miles over uneven ground, should at the last moment have stopped and turned away when only a few yards short of their declared destination is strange indeed. Rightly, it prompts suspicion that some extraneous, hidden factor was operating on the fringes of the incident. But that factor was not the one perceived by Langham and Spencer— that is, a cunning attempt to mislead.

Was Keith a dullard or a simpleton? He would have to be one or the other to believe that the entry he made in his diary about his January 4 trip to Piltdown would convey a true picture of a man befuddled over his whereabouts. After describing his wanderings—occupying several hours of trudging across open land between Fletching and Newick, beside rivers and through valleys, his way at length blocked by bad footpaths so that he and his wife had to retrace their steps—need he have admitted to actually reaching and stopping at that "white gate"? Surely to portray himself as coming at last to the precise spot he was supposedly seeking but really wanted to avoid, and for which the local boys had given explicit directions, then not bothering to open the gate and walk a little farther, was not an effective way to establish his prepared story.

Even a hoaxer of ordinary intellect would know enough to terminate his tale of wandering well short of his destination, perhaps showing himself standing forlorn and perplexed in some nondescript field east of Newick. No conceivable reason exists for the plotter, in trying to show that he missed a certain location, to establish his presence within a few paces of that very spot.

The obvious truth about the diary entry is just the opposite of Spencer's interpretation. Keith and his wife *did* walk through the white gate leading to the avenue of fir trees. They *did* reach the pit. What they did *not* find was the "gravel bed" itself within the pit. The reason for the confusion lies in Keith's customary carelessness in writing his diary entries, a fact readily perceived in their jumbled, incomplete style and racing hand. If the passage in question is isolated from its context, it becomes clearer what Keith in his haste has accidentally left out:

> boys told us where Sussex skull found: fir avenue leading to farm— white gate: on Delta plateau above the Ooze. Didn't see the gravel bed anywhere.

When read at a normal pace, and with no intention to analyze them, as they stand the words convey the definite impression that the writer did in fact go through the white gate. Between the two words *Ooze* and *Didn't* something has been inadvertently omitted, some short phrase stating the act of entry, perhaps a brief *went in* or something similar. What makes this suggestion plausible, in fact nearly certain, is Keith's avoidance of the word *pit*. Like a good geologist, he specifies the gravel *bed* itself as being missed—that is, the bed contained in the pit's lower level.

It was winter. The month was January. Frequent rains, as in every winter, kept the ground soggy and repeatedly filled the pit with water. (Rain caused a long delay at the start of excavations in the spring of 1912.) When Keith writes that he "didn't see the gravel bed anywhere," he is in reality picturing himself, with his wife beside him, standing frustrated at the edge of the *flooded* pit.

Is there any actual evidence that the pit was so flooded on the day of his visit, January 4, 1913? There is. It occurs in the anonymous article, already identified as written by Keith, published in *The Sphere* on January 18, 1913. As he gazed down into the "little trench" the article's author noted that "the pit is now full of water, owing to the heavy rains." An accompanying photograph shows the water level in the pit rising almost to the top.

It is even possible to suggest just how Keith slipped while making that crucial diary entry—that is, why he overlooked the need for some such phrase as "went in." In Keith's busy mind its place was usurped by the words "on Delta plateau above the Ooze." Of course this observation concerning the terrain was not part of the boys' directions, as it may seem to be when the passage is read unthinkingly. It is Keith's own record of the countryside in the region of Barkham Manor, his own professional reaction to the area's geology (in the diary the word "Delta" is indicated by the usual sign, Δ). In noting the nature of the plateau he lost his train of thought. Not realizing his omission, he skips over the phrase needed to denote his passing through the gate and his short walk along the avenue of fir trees. He states only what he saw, or failed to see, after reaching the flooded pit. Succinctly he notes that the *gravel bed* was lost to view.

Arthur Keith, it now may be agreed, was neither a dunce nor a dullard, certainly not if he was in fact the ingenious Piltdown forger. Yet in writing his advance story of the December 18 meeting at Burlington House for the *British Medical Journal,* Langham and Spencer insist, he foolishly gave the game away by using information that, if he were innocent, he could not have possessed. Writing two full days ahead, he supposedly included knowledge that was not made public until the night of the meeting, and some of it not even then.

At first glance this seems like a revelation, may appear well worth serious consideration. But under close scrutiny, this charge, too, readily melts away. In fact the response can be stated in very brief fashion, bare of analysis.

There are two ways, each of them quite innocent, by which the privileged information contained in the article could have reached print. No doubt in this case both were employed. First, the amount of information was by no means voluminous. In conversation with Woodward at the Natural History Museum, where Keith in December 1912 went twice, on the second and the fifteenth, the whole of it could have been conveyed in a sentence or two. Maps and guidebooks would have supplied anything else that was needed. (Spencer's sole argument against accepting Woodward as Keith's source is peculiar, to say the least, and scarcely deserves comment. Keith, says Spencer, "was not a member of Woodward's inner circle of friends.")

Further, anything Keith did not get from Woodward at the museum could easily have come from other informed sources, during or immediately after the meeting on the eighteenth (Lankester, Dawkins, Moir, Barlow, Bather, Dawson himself), or from the newspapers in the next day or two. The article did not appear in the *British Medical Journal* until Saturday, December 21. Up until late Friday night, the twentieth, additional copy could easily have been inserted in the original text. The technique for coverage of late-breaking stories in weekly magazines even then would have called for advance copy to be set in type, then held open until the last moment. After the meeting on the eighteenth, Keith had two days and two nights to make alterations. The *Journal* building stood only a block or two from his office at the Royal College of Surgeons.

In his massive 1992 *Current Anthropology* treatment of the Piltdown affair, Phillip Tobias shows that, like so many other Piltdown commentators, he has failed to grasp the essential point about the *British Medical Journal* article. Like Spencer, he is surprised that someone should have written about an event before it happened. He seems unaware that out of the whole article only two paragraphs are in question, and within those paragraphs only a few phrases. In painful detail, Tobias rehearses what was or was not said at the meeting on the eighteenth about the Piltdown site, at the same time overlooking or denying all other possible sources. He claims to have "pinpointed several items" about the site in Keith's *Journal* story that could not be accounted for legitimately, where in reality these items could have come from any number of people before, during, or after the meeting. Then he makes bold to pronounce a judgment on the two principals, a judgment that is in no way earned. "Collusion between Keith and Dawson," asserts Tobias without the slightest grounding in fact, "may be inferred."

Tobias also strenuously upholds the charge that Keith deliberately made a secret about the start of his friendship with Dawson, hiding the fact that it began no later than—and probably well before—the summer of 1911, at the time of the British Museum's convention. "Keith had taken part," he writes, "in an excursion as a guest of honor to Hastings, hosted by W. R. Butterfield, Dawson, and Lewis Abbott: it is inconceivable that Keith, one of the three guests of honor, and Dawson, one of the three hosts, would have failed to meet." Here Tobias is relying on Spencer's research in the *Hastings and St. Leonards Weekly Mail & Times*, which gives an account of the 1911 excursion from Brighton to Hastings, naming Keith and Dawson among others. For Spencer as well as Tobias, interpreting the details in this paper, there "could be no question" but that a face-to-face meeting, a portentous personal encounter, had in fact occurred between the two men that day.

Once more, however, that sweeping conclusion has not been earned. Two other Sussex newspapers, dailies, give a very different impression of the day's excursion, making it much more of a mob scene. In Hastings during those few crowded hours Keith and Dawson may never have come within fifty yards of each other (if Keith, in fact, actually made the trip from Brighton to Hastings. Not everyone did, and Keith's diary mentions only Brighton).

Dawson was *not* one of the event's three official hosts. Keith was *not* one of only three guests of honor. Descending on Hastings that day were some two hundred convention delegates and their guests. But regarding a controverted issue such as this, where so much depends on the ephemera gathered by reporters, paraphrase or brief quotation will not serve. Here is how the Sussex *Argus* described the day's festivities at Hastings, beginning with the arrival at 10:15 A.M. of the train from Brighton:

> The fascinating romance of Hastings' life-history was presented in cameo yesterday on the occasion of a visit to the premier Cinque Port of the delegates who have been attending the annual Congress of the Museums Association at Brighton during the week. A more delightful method of imparting the essential facts of a town's varied and wonderful career than that adopted was never conceived . . .

The party included many local natural historians, archaeologists, botanists, and scientists—including Mr. T. Parkin, J.P., Councillor Dr. Gray, J.P., Dr. F. Bagshawe, J.P., Councillor Ben Womersley, Mr. C. Dawson, Mr. W. J. Lewis Abbott, Mr. A. Belt, Mr. J. E. Ray, and Mr. Ruskin Butterfield—and throughout the peregrinations through Hastings the members were taken in hand, as it were, by those thoroughly competent to give pithily, brightly, and accurately those facts which link the town up with the time of Athelstan.

Upon their arrival, the delegates proceeded to the museum where they were received by Dr. Gray, Mr. T. Parkin, and Mr. Butterfield . . . The contents of the museum have probably never had so enthusiastic or so numerous an audience. For a full hour they were examined and criticized and inspected, and peered at through magnifying lenses . . .

From the museum they proceeded to the Castle, and under the efficient conductorship of Mr. Dawson and Mr. Lewis Abbott they inspected the ruins and the interesting sites where history was made. Tabloid lectures were delivered and innumerable questions asked and answered, so that by one o'clock the delegates had heard just sufficient of history and of the past to enjoy the luncheon at the Albany Hotel . . .

With various toasts and speeches, lunch lasted for perhaps two hours. The party was then loaded aboard "four motor coaches" and sent on a tour of historical sites in the surrounding countryside. It returned to Hastings in time to catch the 7 P.M. train for Brighton.

The *Hastings Observer* gives an alphabetical listing of the convention delegates and their guests, occupying some six inches of infinitesimal type. Halfway down occurs a simple listing of "Mr. A. Keith, Royal College of Surgeons." Dawson is named by the *Observer* as among the dozen men who were on hand to greet the two hundred delegates, his sole active role being confined to lecturing at the castle. The paper's picture of the busy hour the delegates spent in and around the storied old pile is slightly fuller than that printed in the *Argus*. Here Abbott gets top billing:

Before entering the Castle, Mr. Lewis Abbott led the party in the rock shelters and made some interesting observations with regard to the prehistoric relics of the neighborhood. His description was really the result of his own excavations extending over a period of twenty years, and the relics he had discovered, which illustrated life in prehistoric days, are the most complete record ever discovered in the world.

From Mr. Lewis Abbott's guidance, the party were then led first outside the Castle by Mr. Charles Dawson, F.G.S., F.S.A., who explained to a very interested party the romantic history of the Castle. Then inside, from the position upon which the old Norman keep once stood, he indulged in the reminiscent glories of the Castle's past, of the various occasions upon which it had figured in the nation's annals, and also dealt with the changes it had undergone since the time of William the Conquerer.

Just before half-past twelve the company left this spot, and explored the Castle dungeons, most of the ladies and gentlemen wandering in and about the passages carrying lighted candles, and resembling a great ritualistic procession . . .

A return to town was made by one o'clock for luncheon at the Albany Hotel . . . His Worship [the mayor of Hastings], wearing his chain of office, received the delegates, who were accompanied by many visitors. In all close upon 200 sat down to lunch . . .

This description of the rapid, even whirlwind progress of the delegates among the antiquities of the Hastings area—despite the relatively calmer interlude in the banquet room of the Albany Hotel—hardly encourages the belief that Keith and Dawson "certainly" met and struck up a fateful friendship that day. In truth nothing definite on the point can be stated either way. Objectively no one can say that they did not meet, at least to nod and shake hands. But equally, Spencer, Tobias, and their followers can no longer insist that they did.

Over and over prior to the exposure, Keith praised Dawson as honest and able, having a free and open nature, a man whom he along with many others loved and respected. In the circumstances such expressions of admiration seem no more than natural and expected. Not so, declares Tobias, who insists that these frequent protestations provide damning evidence of the "pattern" of Keith's guilt. His constant support of his confederate, runs the charge, was all too obviously part of an effort to prevent and forestall any questioning of the spurious Piltdown remains.

Keith also spent much time asserting the authenticity of the Piltdown specimens, being frequently "at pains" to defend one or another of the discoveries as undoubtedly genuine, even when no one had raised a question. Here again, says Tobias, is evidence that points unerringly to Keith's guilt: this continual emphasis on authenticity was also meant to ward off any possible suspicion.

These two items of so-called evidence were offered quite seriously by Tobias and have not been distorted or misstated in the telling here. They stand as presented by Tobias in his *Current Anthropology* article of 1992. In discussing them, he agrees that taken separately, they might be quite "innocent," each a normal expression of earnest belief. But when viewed in the general framework of the fraud—that is, when evaluated, as Tobias expresses it, "in the context of all the other lines of evidence"—grave doubts are raised. Confidently he lists these two items among his "Nine Pointers to Keith's Guilt."

Here, surely, is the ultimate illustration of how far from reality, from common sense, the Piltdown investigation has drifted, how far it now stands from any clear idea of what constitutes actual evidence. Perhaps nothing could better demonstrate that deficiency than the fact that in his remark about "other lines" of evidence Tobias has unwittingly mirrored the classic instance of defective proof, almost using the very words. Iago, telling Othello how his wife's handkerchief was seen in another man's possession, says insinuatingly, "It speaks against her with the other proofs." But of course, as Iago knows, there are no other proofs, nor is the handkerchief real evidence.

If Keith knew anything about the fraud before the exposure—and it prompts pity for the aging scientist to think so—he learned of it in 1950 from the publication of Oakley's first fluorine test results. In that year, as a result of the tests, Piltdown's status and high antiquity were seriously questioned. Many scientists proceeded to dismiss it as comparatively recent and of small evolutionary worth, in reality a mistake, a monstrous throwback. For Joseph Weiner, the new information sparked a train of thought that led him to suspect, and then to demonstrate, fraud. For Keith, the fluorine tests could have set off exactly the same train of thought, ending in the same devastating suspicion.

Keith's *Autobiography* was finished in 1949 and was published the next year. In it he speaks with great fondness of his Piltdown involvement. He describes how he afterward persisted in close study of the perplexing skull, and how he firmly believed Piltdown Man to be one of "the earliest known representatives of man in western Europe" (the other being Swanscombe). Obviously, and to Keith's great misfortune, the *Autobiography* was in print and published just before the news of Oakley's disturbing reevaluation broke. Had he known in time about Oakley's results, he might have withdrawn the book for changes, or until Piltdown's status could be settled.

When Keith spoke with Weiner and Oakley at his home in 1953 about the Piltdown exposure, he told them that his correspondence with Dawson had all been destroyed "in a bonfire some years ago." This frank admission calls up a last picture of Sir Arthur Keith, a touching vignette that may stand as the most appropriate ending to his story. Imagine him in 1950, nearing the age of ninety, sitting before his fireplace. Sadly, he reads the letters he had received from his friend Dawson some forty years before, a man for whom he had the "greatest reverence." As he finishes each letter he drops it disdainfully into the flames. Three years later, listening to Weiner and Oakley explain the full extent of the fraud, he could not have been greatly surprised.

The last of the "proofs" in the Langham-Spencer-Tobias theory, the curious matter of the thirteenth vertebra and Dawson's paper on it, at this point hardly needs confuting. Properly, it is not evidence about Piltdown at all, and surely it is egregiously wrongheaded to suggest it as a cover for visits by Dawson to Keith's office. If the two, as charged, were collaborating on a great scientific fraud, their meeting openly *under any conditions* would have been unwise, indeed foolhardy. Not *why* they met, but that they *did* meet repeatedly would have become the significant fact. Dawson's sudden interest in an abstruse anatomical problem, one calling for specialized knowledge apparently beyond his competence, could hardly have deceived anyone if meant as "camouflage."

Here, in reality, is only another instance of Dawson's taking quick advantage of another man's work. He was actually plagiarizing the book by the Frenchman Le Double, newly published in 1912, and covering the very topic of the extra vertebra (Dawson was fluent in French). In asking Woodward to arrange for his paper's presentation to the Royal Society, he emphasized the need for haste: "I am very anxious to get it placed at once . . . I want to secure the priority to which I am entitled." Clearly it was not scientific priority driving Dawson in this instance, but a wish to anticipate an English translation of the French text, and therefore wider general knowledge of the topic of the thirteenth vertebra. The sheer audacity of the move, Dawson well knew, was the very thing that made it likely to succeed.

Sir Arthur Keith was not the Piltdown mastermind. As now will be demonstrated, Charles Dawson in planning and executing the Piltdown forgery neither had nor needed, nor would ever have wanted, a confederate.

11

The Wizard of Lewes

The ancient cemetery associated with the venerable church of St. John Sub-Castro in Lewes is a classic example of those storied burial grounds once so common in a certain type of romantic fiction. Set among old houses and narrow, winding streets, its high stone walls topped by massive shrubbery, it is hidden from the casual passerby. Only a single entrance is to be found, sunk inconspicuously into the curve of the gray stone wall on the church's right flank. Inside is a random sprawl of weathered gravestones, large and small, ornate and plain, intact and crumbling, many disappearing in the gentle embrace of overgrown bushes or screened by the gnarled trunks and branches of ancient trees.

Somewhere in this tangle, his tombstone lost to sight, rests Charles Dawson. Church records show that he was buried here at his death in 1916, but it could take days of searching through the extensive grounds to locate the grave. In death, the man who became England's most accomplished forger remains as elusive as he was in life.

In the four decades since Joseph Weiner made the first hurried indictment of Dawson as the framer of Piltdown, he has been brought no closer to conviction. Despite all that has been written, very little new has been added to the picture, and no investigator has made public any major new facts dredged from Dawson's hidden past that help determine the portrait of this calculating, fiercely ambitious man. Yet what Weiner provided, though it was more innuendo than evidence, implication rather than outright charge, has been enough. For most observers today, whether he is seen as acting alone or with a cohort, Dawson remains the leading suspect, his guilt exceedingly probable.

It is now time to take the final necessary step, to do what could and should have been done at the start. (How Dawson managed to escape for so long, branded only by supposition, is one of the real curiosities of the Piltdown story.) It is time to reconstruct in something like fullness the arrestingly bold, infinitely strange career of deception pursued during thirty-five years by the affable Mr. Dawson. For a detailed look at that career there exist a dozen inviting points of entry, beginning in the early 1880s. But perhaps most serviceable for present purposes is an event that occurred when Dawson had already been imposing for a number of years on the trust and faith of those around him, the affair of the Beauport statuette. Offering what can be termed a "missing link" in the British iron industry, on a smaller scale it is a curiously exact anticipation of Piltdown.

In the spring of 1893, Charles Dawson, then age twenty-nine, went to the British Museum in London, where he visited the keeper of Roman antiquities, A. W. Franks. Proudly he presented Franks with "a little iron statuette," explaining that it had been dug up by a workman from an old Roman "cinder-heap" (a slag mound from an iron-smelting operation common in ancient Sussex). The find had been made some years before, he explained, in 1877 at Beauport Park, a few miles north of Hastings. Such digging in the old mounds, Franks knew, went on all the time in the never-ending search for "road metal" (hard materials used to make repairs to local roads).

The tiny figure had come into his hands in 1883, Dawson declared, when he bought it directly from the workman who uncovered it. From all the circumstances of the discovery, he was sure that the diminutive figure was authentic Roman work, probably made of cast iron. If it was indeed made of *cast* iron (molten metal poured into a mold) and not simple *wrought* iron (hot metal hammered into shape), it would prove a sensational development, history's earliest example, "in Europe at least," of a supremely important manufacturing process.

On that first visit to the museum, Dawson neglected to explain why he had waited ten years to make this exciting artifact known to the world of science, a point Franks himself overlooked and which was never afterward cleared up. He also had failed to record such essentials as the name of the workman and the precise date and other circumstances of the find. Nor did he venture to say why the workman should have retained the little statue for six years—unusual, since the piece was evidently of some real value. Concerning these points, too, Franks failed to inquire.

Only three inches high, the figurine showed a man frozen in action, legs spread, torso leaning. But it was badly damaged. Missing were most of both arms, as well as the lower portion of the right leg, and in addition its entire surface was deeply corroded. "Much rusted and somewhat imperfect," Franks described it, and in a decidedly "decayed state." It had been unearthed from the old mound at a depth of twenty-seven feet. Underneath, at the bottom of the heap, lay some human bones.

Promptly, Franks conferred with his deputy, C. H. Read, and the two men were soon in agreement that, judged on both style and provenance, the statue could certainly be assigned to the Roman occupation of Britain. By comparing it with a similar small statue they already had in bronze, they were able to go even a step further: it was, they concluded, a miniature replica of a known statue of the classical world, one of the horsemen in a colossal group that stood before the Quirinal in Rome.

Unfortunately the little figure was not made of cast iron, or so it appeared. A metallurgical analysis made at the Royal School of Mines judged the material to be "wrought, malleable iron, a steel-like iron, such as was manufactured in early times." Even so, its great interest and importance was undeniable—"Roman works of art in the round are of the highest rarity," declared Franks—and plans were made for its presentation to the Society of Antiquaries in London.

On May 18, 1893, at a regular meeting of the society, with young Dawson in attendance but unable to speak because he was not yet a member, the Beauport figurine was exhibited to an attentive audience. In the chair was the society's president, A. W. Franks, and supplying the description was C. H. Read, secretary. Tests had ruled out the statue's being made of cast iron, explained Read, and he went on to suggest that the actual wrought iron attested just as well to the figure's great antiquity. "If the figure be not cast," he said reasonably enough, "it can only have been made by hammering and chiseling." Such a process was both slow and laborious, he explained, adding (inadmissibly, as Arthur Woodward twenty years later would conclude about Piltdown), "It is inconceivable that it can have been done at a time when the casting of iron was practiced ... the fact of the statuette being of wrought iron adds strength to the assumption that it is of Roman, rather than of later, date." Interestingly, Franks stated, in Germany there existed a few examples of just such ancient iron statuettes, and one of them "is said to be cast."

Three men joined in the subsequent discussion, and what one of them had to say must have jarred the wary Dawson. A. H. Smith, a fellow of the Society, displayed to the attentive audience a small statue of his own. It was made of bronze, he said, holding it up, and all could see that it was about the same size as Dawson's figurine and of the same design, a horseman in action. He had bought his statue only recently, explained Smith, in France at a place called Orange. The seller had offered it as an authentic piece from ancient Rome, and he had bought it with misgivings. Now, after close examination, he had decided it was not Roman at all, but "modern," a replica perhaps meant to be sold as a souvenir. The report of the meeting in the society's proceedings gives no more information about Smith or his statuette, not even whether it prompted debate. But the existence of what was clearly a fake, one resembling the Beauport figure so closely, brought Dawson's presentation to a decidedly lame ending.

The artistic finish of the Beauport statue, a second speaker suggested, its "fine modelling" and the fact that it was a replica of one of the Quirinal figures, were strongly against "its being antique." The last man to offer comment was the elderly and distinguished Sir John Evans, a substantial figure in British science. The similarities between the two pieces was so great, Evans stated, "that suspicion might be aroused as to their belonging to the same category." He was not saying that Dawson's iron statuette was not ancient, but he "preferred to suspend his judgment."

Dawson did not give up. Sometime after the antiquaries meeting he persuaded a government scientist to conduct another analysis of the material in his statue. This time for some unknown reason the test gave the opposite result, and the figure was certified as indeed cast iron, as Dawson had guessed. Then he put his little statue aside for no less than another ten years (at least there is no record of him making use of it in that period). At last in 1903 it suddenly reappeared, included as a prize item in an antiquarian exhibition at Lewes mounted by the Sussex Archaeological Society. In the exhibition's catalog the Beauport statuette stands as the lead item, one of more than thirty objects of iron loaned by Dawson for the occasion. "Dr. Kelner, Analyst of the Royal Arsenal," declares the catalog, "certifies it to be of cast-iron. It is probably, therefore, the earliest specimen of *cast*-iron known."

In an article about the exhibition Dawson is found suddenly providing details about the statue and its original discovery, which he had failed to mention to A. W. Franks at the first presentation years before:

> If we may speculate upon the discovery of one isolated specimen, it would seem that the Romans, or Romano-British, who smelted the iron at Beauport, near Hastings, had already attained the art of casting iron to a great degree of perfection. The specimen referred to was found by one of the workmen employed in digging the iron slag for road-metal about the year 1877. His name is William Merritt, and he lives at Kent Street, Seddlecombe Road, Westfield.
>
> All the workmen engaged in digging were in the habit of picking up any of the more important specimens, such as bastard Samian ware, coins, etc., such as Mr. Rock describes, and keeping them for certain people . . . The author, who had been recommended in the year 1883 to see Mr. Merritt about some geological specimens, procured from him, with other specimens, a small, much-corroded statuette, all of which he stated that he had dug up in the slag-heaps of Beauport . . .
>
> The author, as far as possible, took considerable trouble to settle the question of the *bona fides* of the discovery, and received from Mr. Merritt a written account authenticating it . . .

By this juncture, of course, the original discovery lay twenty-six years in the past, and nothing further was ever heard of the digger William Merritt of Westfield. The written account Merritt supposedly provided went unread by any other eyes than Dawson's, and to date has not been found.

After 1903 the statuette again faded from public notice, although not from public view. As part of a large and varied collection of antiquarian objects loaned by Dawson to the Hastings Museum, it was long on display. Following Dawson's death in 1916 the entire collection, including the little statue, was bought by the museum, where it went on permanent exhibit, now and then finding mention in print. A book published in 1931 about the old Sussex iron industry, *Wealden Iron* by Ernest

Straker, which soon took a place as a principal authority on the topic, calls it "the most interesting object" found at Beauport. But Straker quickly adds a warning, echoing what was said at the 1893 meeting:

> Notwithstanding Mr. Dawson's belief in the authenticity of this find, there are some doubts on the matter. The sale of the objects found was a valuable source of income to the diggers, and it is possible that deception may have been practiced. From the context (of Dawson's 1903 article) it is evident that similar bronze figures have been produced, and a replica in modern cast iron would not be difficult to cast and to corrode by burial.

That is the extent of the public record on the Beauport statuette. It is not, however, the end of the story. A private letter written by Dawson four months after the original 1893 meeting of the Antiquarian Society sheds a good deal of light on the incident's background. In the letter, sent to W. V. Crake, secretary of the Hastings Museum Association, he offers still more detail about the find itself, and even mentions the workman who found it (but this time not giving the man's name or address, information he put on record only a decade later). After some brief preliminaries the letter continues, bringing up among other things an anticipation of one of Piltdown's most crucial features:

> On Sunday I managed to collect evidence at Westfield about my little cast iron Roman statuette. I found the labourer who dug it up and who says it was with coins of Hadrian's time. The coins found their way into the collection of Mr. Rock—late *Rock* and *Hawkins*—White Rock. Do you think you could discover him. I believe he had many losses after retiring. I should be much indebted if you could—and perhaps he has other local specimens he would cede to the Hastings museum. It was *old* Mr. Rock—Hawkins used to collect also.
>
> They rather disputed my iron statuette at the Society of Antiquaries so they put me on my mettle a bit. Fortunately and almost wonderfully, an iron spearhead cast in a mould as in the bronze age, has been found in an ancient iron pit in the west of England and is now on its way to me. The question is had the ancients the knowledge of *casting* iron. I hope it will be a knockdown blow . . .

Almost startling in its resemblance to Piltdown Two is that casual mention of a supposed cast-iron spearhead, recently discovered, that

Dawson is sure will vindicate the original statuette. Not the least part of the resemblance is its stupefying vagueness, both as to location—"the west of England"—and the complete absence of circumstance: no hint of how Dawson had gotten wind of the spearhead or about how he had convinced its distant owners to part with it. But the resemblance ends there. Unlike Piltdown Two, no spearhead from the west of England, or anywhere else, was ever brought forward by Dawson as the "knockdown blow" in support of an early Roman origin for his little statue. His loan collection at the Hastings Museum contains several such spearheads, one of which may have been intended for this duty. Apparently by then he had learned a lesson invaluable to the forger, the fact that bold assertion often proves sufficient.

More interesting than the spearhead is the letter's mention of "Mr. Rock" and the Roman Emperor Hadrian, for there the usually careful Dawson finally slipped. In 1879 James Rock contributed an article to the annual *Collections* of the Sussex Archaeological Society entitled "Ancient Cinder-Heaps in East Sussex." It was largely from this pioneer article that Dawson first learned of the huge Beauport slag heaps. It was also from this article that he derived his daring idea about a forgery that would establish Sussex as the birthplace of cast iron, with himself in the role of astute discoverer.

Of fair length at fifteen pages, Rock's article treats of several ancient smelting sites in Sussex, but mostly in Beauport Park. Until it was opened up, about 1870, as a quarry for road-mending materials, the mound of *scoriae* at Beauport had stood for centuries concealed under a layer of earth as a grassy hill or knoll supporting numerous trees. Fully two acres in extent, its highest point reached fifty feet, but once uncovered its reduction by the constant work of diggers was swift. By 1879 scarcely a fourth of the original slag heap remained, and several more years of digging, thought Rock, would certainly see it exhausted.

All the evidence pointed to the mound being the detritus of an iron-smelting operation in Roman Britain. Describing its peculiar layered form, Rock tells how the hill slowly accumulated under Roman methods of smelting iron ore, and he ends with a remark about "the molten iron running off from the ore to the bottom of the mound." Molten iron would allow for casting in molds, so a cast-iron statuette, suitably small and convincingly damaged, would seem a reasonable, even likely discovery at Beauport Park.

But Rock was wrong. The primitive methods of smelting practiced in Sussex by the Romans, clearly deducible from the slag heaps themselves, could never have attained a heat intense enough to produce molten, free-flowing iron. The most that the Beauport furnaces accomplished was the elimination of impurities, leaving a spongy mass that rapidly cooled and hardened.

The evidence Rock presents for a Roman date, however, is convincing. Besides pottery, there were two coins, one of Trajan and one of Hadrian, both of which, he adds, "I have in my possession." These yielded a secure date of the mid-third century. In his letter to Crake Dawson states that according to the workman the statuette was found "with coins of Hadrian's time" (the oversight about Trajan is not material). This, of course, would be an extremely important fact for Rock to have recorded, for if he had known of the little figure's existence he would certainly have featured it as his greatest prize. Yet in his article there is no mention of anything resembling a piece of sculpture. James Rock, the leading authority on Beauport, writing within a year of the supposed discovery, never heard of Dawson's statuette.

Similarly, Dawson claims that in 1883 he bought the statuette directly from the workman who had found it six years before. But the workman in selling the object to Dawson, or to anyone else, surely would have also offered the coins. Even if he had already sold them, he would certainly have mentioned the fact of their existence, thus boosting his price for the statue. Yet ten years later Dawson showed no awareness of the coins, and at the 1893 antiquaries meeting nothing was said about them. It was only four months after that meeting, in a private letter to a museum official, that Dawson thought to mention the crucial fact that the coins and statuette had been found in association. He had learned this, he says, only days before—"on Sunday"—from the workman himself.

In his anxiety to rescue the bogus figurine from the negative response of Sir John Evans and the others at that meeting, Dawson unwittingly created a peculiar contradiction. He has the Beauport workman (at this point still unnamed) twice acting against his own interests by withholding critical information: when selling the coins to Rock he kept silent about the statuette, and when selling the statuette to Dawson he kept silent about the coins. Any worker accustomed to digging in British soil would have understood that selling statue and coins together would have brought the best return. But if for some reason they were to be sold separately, then informing each buyer of the impressive dual find would have been both natural and imperative. Then, by 1903, the coins were again forgotten. In the article Dawson wrote for the Lewes exhibition that year, though they would have greatly strengthened his claim, they receive no mention.

It was in this same 1903 article that Dawson first publicly named his Beauport workman: William Merritt of Westfield (a nearby town), even supplying an exact address. Then, rather unnecessarily, he went a step further. At his request, he says pointedly, Merritt took the trouble to write down a full description of the actual discovery. As Dawson expected, that claim proved to be both effective and safe. There is no later record about the workman Merritt, or about anyone bothering to visit and speak with him. No doubt the man existed. A check of Sussex obituaries, say for 1902, a year or so before Dawson named him, perhaps would reveal the recent passing of at least one Mr. Merritt. But in this case Dawson may not have been even that careful. If Merritt was ever sought, and not found, by some investigator—a very unlikely action in such matters, in those days hardly ever pursued—Dawson could easily claim a mix-up of name or address, or confusion in any number of innocent details.

Usually it is impossible to say how or why the precise idea for a particular fraud or hoax is born. In the present case, however, it is probable that some passing remarks by James Rock in his article supplied the trigger. At Beauport Park, he states, definite proof was found of "Roman iron manufacture in a district which has hitherto supplied but little evidence of the kind, and is but little known." Not only smelting but manufacture was practiced at Beauport by the Romans, says Rock, and there was also his prior (and mistaken) reference to molten iron running off from the heated ore. Best of all, about the site itself there was "little known," an absence of specific knowledge that offered fertile ground for an imposture. If in about 1891 Dawson was looking around for another subject on which to exercise his considerable daring, Rock's words would have seemed an open invitation.

His scheme determined, Dawson needed only to obtain an appropriate modern replica of an ancient statue, such as the bronze copy bought by Smith in France and displayed at the antiquaries meeting. Preparing an actual mold from the replica, then casting a new example to be suitably corroded, may have proved too troublesome. Perhaps he simply doctored the replica.

During the final decade of the nineteenth century, Dawson perpetrated half a dozen or more frauds, none quite as elaborate as Beauport, but all in their own way ingenious. Revealingly, as Weiner was first to note, though without reference to fraud, they hovered around the topic of "transitional" forms or missing links. Two of them, a flint weapon and an ancient boat, were accorded official publication. The third, a unique horseshoe, was successfully submitted to the Society of Antiquaries. A brief description of all three will show Dawson at his most resourceful, boldly intruding himself into areas where it would have seemed he had neither training nor competence.

In the *Sussex Archaeological Collections* for 1894 there appeared an account of the simplest of the three, briefly described as a "Neolithic Flint Weapon in a Wooden Haft." It is short, occupying only three pages, one of which is given to illustration. The weapon itself, it seems, at that time no longer existed, at least not the wooden handle. When it was originally uncovered, three feet down in the soil, the two-foot-long handle had been lost almost immediately. As Dawson explained, it "was perfectly carbonized and crumbled at the touch, and all attempts to save it proved futile." Surprisingly, it was not Dawson who made the discovery, and it had not happened recently, but "some years ago." An amateur collector of flints named Stephen Blackmore, of Eastbourne, a man who made his living as a shepherd, had come across the weapon while digging on the sea cliffs near his home.

It was sometime in 1893 that Dawson visited Blackmore, assigned by the Hastings Museum to purchase a portion of his extensive flint collection (so much is true). While in the man's home, said Dawson, he noticed "a drawing of a haft bearing an instrument *in situ*." Asking about it, he was told of the discovery on the cliffs, and of the failed attempt to save what Blackmore took to be a disintegrating handle. The elderly shepherd, recounted Dawson, showed himself to be a "fair draughtsman," so at his visitor's urging he consented to make another, more detailed drawing of the weapon as it had looked when lying in the ground. As he worked, the two men talked, deciding on such details as measurements and thicknesses, and how the flint blade was lashed to the handle. The blade, they agreed, had been inclined slightly downward, and in the handle there had been a gentle bend or back curve. Both of these were characteristics that "might be noticed in the construction of our own modern implements."

Included in the drawing is the flint blade, but it escapes notice in the text. Since Dawson says nothing about handling or inspecting the blade, perhaps it too was drawn from memory. That he suggested or added a touch here and there to Blackmore's drawing may be taken for granted, small touches, no doubt, but transforming in their effects.

After publication of the article no more was heard of Blackmore and his hafted flint. But the authoritative report, with the sketch of the weapon carefully attributed to the shepherd, added its bit to Dawson's growing reputation as an antiquarian, one always keen to spot what others had missed.

The boat came next. It was described in another article in the *Sussex Archaeological Collections* that same year, 1894, and involved even more subtle manipulation than the flint weapon. Its circumstances dated to 1887, and concerned the remains of a small vessel, like a rowboat, found buried on the beach at Bexhill. Dawson involved himself with the find because, as he said, it furnished "an interesting link in the history of boat building." To him it seemed to provide a transitional phase "between the coracle and 'burnt out' boat, with the more modern type depicted in the Bayeux Tapestry." When the broken and decayed parts had been reassembled by Dawson, the resulting boat stretched nine feet in length, with a six-foot beam, an eighteen-inch depth, and a flat bottom with prows fore and aft.

As Dawson explained, the buried remnants of this boat had from time to time over many years lain partially exposed on the sand. But only in 1887 did someone, a local boatwright named Jesse Young, venture to salvage the pieces. It was shortly afterward that Dawson came across the fragmented relic lying in a heap "outside Mr. Young's shop." Any full-scale restoration, he quickly saw, was out of the question. But he made an "exhaustive examination" on the spot, and with Young's assistance was able to arrange the pieces "in juxtaposition." He then drew a sketch, which was used to illustrate the article. The text supplied a detailed description of the boat's planking, fastenings, and general shape.

It was all very professionally done, and the article in the *Collections* must have greatly enhanced Dawson's standing in archeological circles. The only disturbing fact is that at about this same time a local newspaper also carried a report of a find made on the Bexhill beach, involving a small boat very similar to that of Dawson.

The account in the *Southern Weekly News*, of January 21, 1888, specifies that the discoverer of the find "recently made at Bexhill" was one Mr. Webb, foreman of a construction crew at work on a park nearby. Embedded in the sand was "what appeared to be an old boat," and Mr. Webb promptly set his men to work "with pick-axe and spade" to dig it out. This haste was unfortunate, complained the paper, for the relic was freed from the sand "in detached pieces instead of intact. The timbers being too rotten it was found impossible subsequently to readjust them, even if the ingenuity of a carpenter would have been equal to the task." Though the paper rightly lamented this careless approach, it had given Dawson his opportunity.

Discarded by Webb, the jumble of rotting oak was carted by Jesse Young back to his shop, no doubt a labor of love for the professional boat-builder. When Dawson soon after encountered the wooden tangle on the grass in front of Young's shop he quickly realized that the remaining pieces could be "juxtaposed" in several different ways, none of which could be seriously challenged. The shape he envisioned, inevitably, was something unique, something that would fit the gap between old coracle construction and the more modern frame-and-plank style. The watching Jesse Young, no doubt, was as fascinated by the suggestion as were the readers of the *Collections*.

The unusual horseshoe came last, this one found, he claimed, by Dawson himself. The site he chose for his discovery was also unusual, as well as imaginative and a little daring: his own hometown of Uckfield. Through one end of town flowed the Ouse River, which was crossed by a bridge at a main thoroughfare. Still in place in the river bottom, explained Dawson, were the piles of a previous bridge far more ancient, and it was from among these rotting old piles that he fished his "Roman horseshoe." Made of iron, it had nail holes dotted evenly around the front edge. But on the sides and at the rear it had only a "flange terminating in a hooked-shaped piece as if used to strap the hinder part of the shoe to the horse's hoof."

In ancient Rome, according to classical historians, horses never had shoes nailed on. The animals were either left unshod or had tied to the hoof a sort of rough slipper of metal or leather. The nailing of shoes directly to the horses' hooves came somewhat later, and not in Rome but in Gaul.

Dawson's find at the Uckfield bridge—made within two blocks of his residence at No. 1 Aylesford Terrace, a two-story private home he rented—provided a transitional stage in the horseshoe art, one never seen before. As such it was readily accepted by Sir A. W. Franks of the London Antiquaries, who called it a "development" of the Roman practice. Obviously Franks never paused to ask whether a transitional phase in horseshoe technology made any real sense: if the front of the shoe was better secured by nailing, why not the sides and back as well? Why continue to tie one part? All such questions went unanswered.

The last that was seen of the Uckfield horseshoe was at the grand antiquarian exhibition of 1903 in Lewes, where it was featured. But before disappearing it too increased its finder's reputation as a man whose restless mind and perceptive intellect missed little of antiquarian worth in his native Sussex.

So far as may now be determined, Dawson's manufacture of fraudulent objects, mostly of the inconspicuous sort, went on up to about 1895. His operations for the most part embraced metal items, but included also were such unusual objects as a clock face from the Middle Ages bearing a painting of human figures, and a small mace that had been carried centuries before by a Hastings "water-bailiff." Both of these items enjoyed fairly long lives as authentic rarities, but eventually it was realized that the clothing on the clock figures was anachronistic, and the mace, revealing itself as dating from the nineteenth century, proved unconnected to any water-bailiff.

During this time several small frauds worked in iron seem to have misfired. There was a diminutive anvil bearing the date 1515 (the stamping of the date was much too clumsy and the form of the letters was anachronistic) and a miniature ax head whose blade measured less than two inches across (its shape seemed medieval, said one expert, but he did not know of another so small). There was also a prick-spur supposedly used by a Norman horseman, but it elicited little or no interest. ("It is certainly not a prick-spur, and is nothing we can recognize," replied the British Museum when queried in 1954.)

As early as the fall of 1890, when he was twenty-six, Dawson was garnering publicity for his discoveries. The *Southern Weekly News* of November 15 that year reported that he had found in the ruins of Hastings Castle a rare circular bronze seal. It was two inches across and bore a heraldic coat of arms linking it to the ancient Abbotts of Battle. Also in Dawson's possession at the time, said the paper, were several other interesting objects of bronze, mainly two unusual ax heads and a dagger hilt of peculiar design. All had been found at the castle, continued the paper, and would shortly be reported and described at a meeting of the Sussex Archaeological Society. But nothing further was ever heard of these particular bronze items. (The frugal Dawson held on to them, however, for they show up in the inventory of his loan collection at the Hastings Museum.)

Precisely what use Dawson may have made of these small frauds is lost to history. Each in its turn, of course, would have been produced with some definite effect or target in mind—impressing some highly placed museum official or an influential officer of some scientific body. There is no evidence that any of these fabrications ever brought Dawson a monetary return. In fact, his earnings as a solicitor must often have been stretched to cover his obscurer activities, especially when he stepped up production or risked working on a more ambitious scale, for instance with Piltdown. At this point, in all justice, must be quoted the single sentence of personal detail to be preserved about Dawson's wife, the widow Postlethwaite. The words were uttered by Maude Woodward, wife of Sir Arthur, not out of ill feeling but as mere information: "Mrs. Dawson was rather a plain woman, but she brought some money to the marriage, I believe."

So far untouched by suspicion of fraud is the earliest part of Dawson's career, the period prior to 1884, when he was wholly concerned with fossils. In that year his collection of dinosaur, reptilian, and mammalian bones, gathered in the Weald, was purchased without hesitation by the Museum of Natural History. Not a breath of concern about those specimens arose then, and none has surfaced since. No doubt it was a result of the high standing Dawson earned as a result of that transaction—accomplished when he was only twenty years old—that disposed Arthur Woodward in November 1891 to report on Dawson's latest fossil find to the Zoological Society of London.

Woodward, the same age as Dawson, was then an assistant at the Natural History Museum. Just how he and Dawson were led to collaborate on *Plagiaulax*, a supposedly new—in fact transitional—form between reptile and mammal, cannot be said. In any case, looking back now from the vantage point of the full disclosure of Piltdown and the other frauds, and conscious of the cloud that must hover over everything that Dawson touched, it may be guessed that here, too, resides an imposture. All too quickly that guess finds strong support.

The identification of *Plagiaulax* rested on a single tiny tooth. As so often with Dawson's later finds, the date and location of the discovery are both left vague, listed only as sometime in early 1891, in a quarry near Hastings. Supposedly the tooth was found embedded in rock, and when Dawson presented it to Woodward it still had a bit of its hard matrix clinging so tenaciously that the museum's lab technician could not dislodge it without harm to the specimen. The crown was greatly worn and much of the enamel had disappeared. Of the cusp, only a portion could be seen, as also the roots.

After careful study, Woodward found that though its configuration was different and it was definitely larger, it bore an undeniable resemblance to some other Mesozoic animal teeth, those of the multituberculates. With this, he added a comment that sounds almost startling in its premonitory echo of Piltdown:

> It only seems to differ in the extraordinary amount of wear to which the crown has been subjected, and in the appearance of this abrasion not having been produced entirely by an upward and downward or anterior-posterior motion, of which the jaws of the known Multituberculata seem to have been alone capable.

The tooth's wear pattern, Woodward conceded, was all that distinguished it from similar ancient mammals already known. Yet he didn't hesitate to reach a fairly radical decision. Based on the tooth's size and configuration, he concluded, it represented a new species in the genus *Plagiaulax*. "Until the acquisition of further material, the species in question may bear the provisional name of *Plagiaulax dawsoni* in honor of its discoverer."

Dawson, it can be said with certainty, was not unaware of how pliant and accommodating his earnest friend could be in such matters. Twenty-one years later, under Dawson's careful manipulation, that early blunder concerning dental-wear patterns would be repeated by Woodward on a far larger and more significant scale.

At a meeting of the Antiquaries Society of London in the fall of 1895 Charles Dawson was elected a fellow. It marked his second election to a prestigious scientific body, his initial such honor having come from the Geological Society ten years before. At the comparatively young age of thirty he was entitled to have the initials F.G.S., F.S.A. appended to his name, a satisfying accomplishment for a man at any age, especially so for one who was without a university degree. But as the evidence indicates, Dawson still was not satisfied. He had now set his sights on the ultimate prize in the intellectual arena, a fellowship in the Royal Society, and for this campaign he changed his tactics. A reputation as a historian of British antiquities was his new goal. To achieve it, he turned his back on the fabrication of actual objects and took up plagiarism.

The first traceable instance of Dawson stealing other men's work occurs in an article of his published in July 1898 in the *Geological Magazine:* "Ancient and Modern 'Dene Holes' and Their Makers." This concerned an old topic of Sussex history, a random series of caverns spread throughout the county, as well as adjacent counties, all sunk deep in the earth and reachable only by way of vertical shafts. For years debate had centered on the question of their origin and purpose, and in Dawson's time the controversy was still very much alive. His paper, a fairly lengthy one, covered the whole range of the topic and suggested some interesting new ideas.

No sooner did it appear, however, than another article on the same subject was submitted to the magazine and was published that October. Written by a geologist of nearby Essex, T. V. Holmes, it commented on Dawson's theories, continuing the old debate, but in his opening paragraphs Holmes also included a statement that came very close to charging Dawson with theft.

More than a decade before, said Holmes, his organization, the Essex Field Club, had issued a report on local dene holes. In his article Dawson had referred to Essex, so Holmes asked space for a reply:

> . . . our views do not appear to be injuriously affected by Mr. Dawson's remarks, yet as the dene report is now more than ten years old it is probable that few of the readers of the *Geological Magazine* have both seen and remember it. And the impression that the reader would derive from Mr. Dawson's article is that his bell-pit hypothesis is something quite new, and therefore uninvoked by us, whereas it was an old view before the report was written, having been put forward by the late Mr. Reave Smith in *The Gentlemen's Magazine* in 1867; and an account of working for chalk of the kind described by Mr. Dawson, written by Mr. F. J. Bennett, of the Geological Survey, is appended to the Dene Report.

The language of the passage is circumspect, yet few readers could have missed its true import. The writer, Holmes, is not arguing for mere priority over Dawson. He intends to cite the sources Dawson used but left unacknowledged.

After the exposure of Piltdown in 1953 some of Dawson's writings, including the paper on dene holes, were studied, privately and informally, by several qualified persons. Dawson on dene holes, it developed, represented a clever pastiche of outright quotes and deft use of borrowed ideas. In the writing, however, he had been able to create a sense of his personal involvement by visiting an actual dene hole and descending into it (or at least claiming he had): "The writer made the usual descent into the pit, which is performed by placing the toe on the hook of the cord and holding the rope above, the windlass being carefully unwound by the man at the surface. With a frayed rope not an inch in diameter this may seem dangerous; but few accidents have been known to occur." By now Dawson had learned the indispensable lesson that a lie, told well and coming from a trusted source, will almost always succeed. But even where a lie is suspected, the teller need not suffer: despite the Holmes article, Dawson's reputation was not affected in the least by his shaky venture among dene holes.

Similar informal studies of a few other Dawson articles were also made at the time of the Piltdown exposure, and the general results were publicly noted, though sketchily. In his 1955 book, Weiner mentions one of Dawson's most important efforts as a historian, a lengthy piece on the history of the iron industry in old Sussex. "The greater part," he states, "has been taken from an early writer, Topley, almost word for word without acknowledgement." But Weiner might have gone much further than this, for in his possession at the time he was writing his book was an exhaustive analysis of Dawson's iron article, prepared for him by an expert in the field, R. L. Downes.

The article in question appeared in the *Sussex Archaeological Collections* in 1903, in connection with the Lewes exhibition of that year. Page by page, Downes traced bits and pieces of Dawson's text to specific authorities who during the previous forty years had written on the Sussex iron industry. Three quarters of the article had been plagiarized, he found, taken directly from at least ten published sources, but mostly from William Topley's *Memoir on the Geology of the Weald* (1875). All the usual tricks of the accomplished plagiarist were readily discovered.

There were specific citations tied to a fact or two, when in reality whole sections had been lifted from the cited work. There was the breaking up and paraphrasing of longer passages taken bodily from the source, with a subtle weaving together of facts and phrases from different texts. There were incomplete or deliberately inaccurate citations. Dawson's own contribution to this mélange amounted to a few simple comments, some personal observations, and transitional material.

Downes also exercised his searching analysis on another Dawson article, "Old Sussex Glass," published in the *Antiquarian* in January 1905. It, too, proved to be largely the result of thievery, in this instance fairly easily traced and showing the same useful pilfering techniques.

Dawson's two-volume *History of Hastings Castle*, it developed unsurprisingly, comprised his boldest and most comprehensive literary theft. At the time of the Piltdown exposure, the book was closely inspected by two scholars separately (Manwaring Baines, curator of the Hastings Museum, and R. L. Downes). Both found that its use of sources, especially an unpublished manuscript by an earlier historian of the castle, William Herbert, involved the same type of manipulation traceable in his articles. "Mr. Baines declares," wrote Weiner in 1955, "that half the material in Dawson's volumes is copied unblushingly from Herbert's manuscript, and describes the rest as gross padding."

With that estimate Downes fully agreed. His opinion of Dawson's text (this opinion was also supplied to Weiner but not quoted by him) was every bit as damning as that of Baines. The compilation of the two volumes, reported Downes, "involved a considerable amount of labour. But this was not a tithe of the labour it appeared to involve. He does not state specifically that he has done all the research himself, but he implies as much in his Preface. The absence of acknowledgement to the sources from which he took his material is scandalous."

Dawson's old trick of appearing to give adequate credit to a source, then using far more material from that source than is admitted, was in full operation with his *History*. In fact, it led several later commentators into a serious disagreement on the point, some insisting that the mention of Herbert in the lengthy preface was sufficient to cover Dawson's use of the Herbert manuscript in the text. But these defenders, though quite sincere, were simply not familiar with the actual sources or with the full text of the two-volume *History*. They were, truth to tell, more of Dawson's dupes, though having even less excuse for being caught than those who welcomed the book originally. In reality, Dawson's preface is a very deft performance in which he presents himself almost as a polyhistor, one, moreover, who is competent in languages ancient and modern, and having almost a professional familiarity with the arcane world of old documents. Of course none of this was true.

Two last articles, widely divergent in subject matter, closed out Dawson's efforts to earn a name as a historian. One is concerned with the famous Bayeux tapestry, and appeared in the *Antiquarian* in 1907. For many years this article was cited in tapestry literature as an important contribution, and since no study of possible sources has yet been made, nothing can be ventured here about its origins. It may in fact represent that extremely scarce commodity, original, underived work by Dawson. After all, there seems no good reason why a forger and plagiarist might not, now and then, have a genuine flash of inspiration. Still, common sense counsels otherwise.

The other article deals with a geological mystery and is definitely not Dawson's work. Similar to the dene holes he treated earlier, this paper concerns the so-called Red Hills of Essex. It appeared in the *Antiquarian* early in 1911 (written perhaps in 1909), offering a supposedly new theory to account for what had long been regarded as artificial mounds of reddish earth. But as with his paper on dene holes, this effort, too, drew an immediate response, again from men who were thoroughly familiar with the specialized topic.

Politely but firmly, their response states that Dawson's ideas had all been anticipated in previous writings on the Red Hills, something that should have been known to anyone in the least acquainted with the subject's "inner workings." Baldly, the response closes by stating that "This summary of the evidence . . . although largely a repetition of what has already appeared in the reports, will perhaps serve more clearly to enable others to judge how far the points which Mr. Dawson has raised are capable of acceptance." It was not acceptance of Dawson's ideas that the writers had in mind, but a charge of plagiarism in those conditions could hardly be made outright.

Dawson's willingness, always rather blithe, to intrude himself into areas where he had no perceptible competence is a conspicuous mark of his character. It argues a high if quirky intelligence, as well as an impressive mental flexibility and quickness, along with an energetic, cramming sort of aptitude for study of the appropriate authorities. In action, the trait may be seen best, perhaps, in his involvement with a natural gas discovery at Heathfield, a town just east of Uckfield. Here no fraud was involved, apparently, and all the circumstances seem normal. Yet Dawson, who did not make the discovery and had no official link to it, ended by gaining most of the credit. In addition, while no fraud can be detected, *something* peculiar seems to have happened.

In boring for water near the Heathfield train station, a railroad crew hit a pocket of natural gas. Deliberately ignited as a test, it sent a plume of flame shooting up some fifteen feet. This happened in the summer of 1896, but for some time afterward the discovery was kept secret. Eventually Dawson heard of it from an eyewitness, and he somehow obtained samples of the gas, which he gave for analysis to his Uckfield friend, the chemist Samuel Woodhead. Informed by Woodhead that the gas was a petroleum derivative—exciting news, raising the possibility that oil underlay the Weald—Dawson promptly sent a report to *Nature*, which was published in the December 16, 1897, issue. Discussing the matter briefly but with an air of learning and expertise, he ends with the assurance, "I am making experiments with the gas, and examining the cores of the boring with a view to ascertain the source of the supply."

In January more tests were conducted on the site by Woodhead, Dawson accompanying him, which confirmed the original hasty lab analysis. In the summer of 1898, Dawson twice went into action, capturing public attention by presenting papers on the discovery to two prestigious scientific bodies. Early in June he spoke to a gathering of the Southeast Union of Scientific Societies at Brighton, then a week later repeated the talk for London's Geological Society. In each case his treatment of the topic was full and confident, covering the chemical nature of the gas as well as its probable origin and predicted volume—all laid out with the air of a first-rank petroleum engineer. The drama of his London performance he enhanced by having an assistant display two large burners on the center table, the flames lit by gas brought in cylinders from Heathfield.

However, this particular evening was not a complete success for Dawson. When he finished his presentation, another speaker arose, a chemist named Dr. John Hewitt. It seemed that a second analysis of the gas, made apparently at the request of the railroad company, had been carried out by Dr. Hewitt. His results, surprisingly, proved to be quite different from those of Woodhead—in fact just the opposite. No petroleum was involved, announced Hewitt, only methane gas arising from decayed vegetation. The contradictions were striking. Where Hewitt detected the presence of hydrogen and nitrogen, Woodhead found nothing. Where Woodhead detected carbon monoxide, Hewitt picked up nothing. Both found methane, but in markedly different concentrations.

In the discussion period Dawson could suggest about these discrepancies only that if both analyses were taken as correct, "then the gas must vary considerably. Although variation had been noticed in the North American natural gas, it had scarcely been to this extent." There the question was left, never to be resolved.

No oil was ever found at Heathfield or anywhere in England's south. The Heathfield gas itself, however, continued to flow and did duty for many years, being piped for lighting purposes to the nearby railway station. Soon, and as it has ever since, the Heathfield gas came to be regarded as Dawson's personal discovery, its harnessing by the railroad all his doing. In Piltdown literature up until the present the curious mistake has persisted.

Because he died so soon after fame sought him out, Dawson's personal life has never been known in anything more than outline. As a result, no one today can presume to say what precise circumstance or event led him, behind a facade of respectability, into a lifetime of fraud. About all that can be suggested is linked with his two brothers: Hugh, a successful clergyman, and Trevor, an even more successful executive and military officer. Both Hugh and Trevor, each some years younger than Charles, had attended the university, while Charles for some unknown reason had not. At age sixteen he went directly from school at Gosport, near Plymouth, to an apprenticeship in a Hastings law firm. This was in 1880, and over the next few years he had to be content with what seemed an unexciting future while watching his two brothers steadily rise in the world.

Trevor's achievement in particular was impressive, for it eventually included entry into London's Mayfair society and a directorship in the important firm of Vickers Ltd., as well as a place on the boards of several other leading companies. In 1909 he was given a knighthood. His later honors, though of no effect in Charles' life, display something of the formidable challenge Trevor posed to his older brother. In World War One he rose to commander in the Royal Navy and was much decorated for his service. In 1920 he was created baronet. *Burke's Peerage* has him receiving the "silver medal of the Royal Society of arts and the Gustave Camet gold medal."

Charles Dawson's darkly dual course in life may have been ruled by a fierce desire to rival and outdo his brothers, especially the younger Trevor. It may not. Until more information surfaces, nothing certain can be known about such matters. Still, the idea of a rivalry does serve to highlight Dawson's immediate purpose with his various frauds, the gaining of public honors by election to scholarly bodies, something neither of his brothers achieved. Twice he succeeded in this design, but then, early in the new century, he found himself stalled. After a decade of chasing what all regarded as the ultimate in such recognition, a fellowship in the Royal Society, he had fallen short, deprived even of the satisfaction of a proposal to that body. But he did not relent. As the evidence reveals, he again switched tactics, after a long lapse once more taking up the forging of physical artifacts. This phase of his lengthy and remarkable career began with the affair of the Pevensey brick.

Resembling his earlier ambitious effort with the Beauport statuette, the Pevensey brick, or stamped tile, presents a wonderfully complicated and daring scheme. Also like Beauport, it was eminently bold, and in a few of its features could be said to have anticipated Piltdown, then just on the verge of being conceived in all its astonishing control and sweep.

The ruins of a large Roman fort had long been known to lie buried in the Pevensey area, on England's south coast halfway between Hastings and Brighton. Here, late in 1906, a full-scale excavation was mounted under the leadership of Louis Salzman of the Sussex Archaeological Society. In the digging many Roman and medieval artifacts were turned up, one of which in particular caused excitement: an ordinary building brick or tile, blue-black in color and bearing a stamped inscription. Not only did the brick identify the site of a long-lost Romano-British town, it also dated the fort itself to a critical period in British history, the withdrawal of the Roman legions in the early fifth century A.D. This ordinary brick rapidly became a physical document of great importance for British historians.

Actually, two bricks were involved, one nearly whole specimen, along with a fragment of another. The smaller of the two, the fragment, was found early in 1907 by Salzman's team after it had been at work for several months. Tantalizingly, its damaged inscription trailed off into the brick's two broken ends, with some of the letters only shadowy remnants: "... ON ... Aug ... NDR ..." Many guesses at the possible meaning of the letters were offered, causing much energetic discussion. But the matter was not resolved until Dawson heard of the find—upon which he came forward with another, more complete example of the same brick.

In its entirety, the words in the abbreviated phrase of Dawson's brick read, "HON AUG ANDRIA," which all promptly recognized as referring to the Emperor Honorius, who reigned during 395–423 A.D. For the experts the final syllable was equally obvious. It provided a clear reference to the important town of Anderida, named in many authentic Roman sources.

This new brick Dawson showed first to Salzman early in 1907, who was delighted to have solid confirmation of his team's find. Then, in April, Dawson made a formal presentation of the brick to a meeting of the Society of Antiquaries, displaying along with it the Salzman fragment. His own discovery, he announced, had not been made recently, but some five years before, in 1902. It had turned up in a pure accident, which might have happened to anyone. The brick exhibited on the table in front of the members, he said, "I was fortunate enough to discover beneath the arch of the postern gate in the north side of the wall . . . It had evidently fallen with other pieces from the roof of the arch, where similarly burnt bricks are to be seen."

With that, he went on to deal learnedly with the name Anderida ("Anderesium, or Andredes-ceaster . . . but although there is a certain general phonetic resemblance between the names according to the way the accents are placed, the philological connection is by no means clear . . ."). Perhaps, he suggested, it was a geographical name adapted for the Pevensey *castra*, or fort. The brick, he said, strongly indicated that the walls of the old fort had been erected, or at least repaired, at a period very close to the start of the fateful Roman withdrawal. His listeners all understood the high value of the stamped bricks, beginning with the fact that these were the only epigraphic occurrences of Honorius' name found to that point in Britain. Curiously—as so often!—in the discussion that followed no one ventured to inquire why Dawson had waited so many years to unveil such an exciting discovery.

Returned to its own museum in Lewes, the Salzman brick was put on exhibit. Thereafter, the provenance of the Dawson specimen seems to have been temporarily lost to sight, eventually showing up in the collections of the British Museum. Meantime, the inscription itself steadily worked its way into the standard histories of Roman Britain, finding a conspicuous place in books and scholarly articles tracing Rome's relinquishment of its outlying territories.

The revelation of fraud came in 1972. The many bricks still in place in the fort's walls, it was found, were quite different in composition from the Dawson-Salzman bricks, and none bore the Honorius stamp. To determine their age an established test was applied, a thermoluminescence measurement. Both bricks proved to be less than a hundred years old, more likely as little as seventy. The Pevensey tiles had been made and fired not much before 1902.

The full implications of the exposure were readily perceived. Dawson had fabricated not only his own brick, but also the one found by the Salzman team (analysis showed that the two inscriptions came from the same die). After making both bricks, he had cunningly broken one of them, leaving only enough of the stamping to puzzle and stimulate its finders. The pity is that no hint remains of how or when Dawson accomplished the planting of the Salzman fragment in the Pevensey fort, who found it and how, and exactly where in the fort it turned up. Nor is there any record of Dawson's visiting Pevensey while the 1907 excavations were in progress. But he was there.

Considering all that is revealed in the preceding extended review of Dawson's varied activities as a forger, both mental and manual, and of his intellectual prowess in general, it is nothing short of silly to maintain that with Piltdown he could have *required* help. Possessed of a mental capacity characterized by unusual breadth and quickness, when his interests were engaged he became a lightning assimilator of knowledge, both as to concept and detail. Within the limited framework of the moment, and under the pressure of necessity and concentrated effort, he could actually become more expert, more proficient for the time needed than the leading authorities on whatever specific topic was in question. All too obviously, the precise information he required for Piltdown—principally in geology, paleontology, anatomy, and chemistry—he could have made his own at will. Here once again, as so often with those who have left a negative mark on society, the old adage applies. Were it not for some wrenching quirk of personality, some twist of fate, Dawson's remarkable talents might well have earned him a respected niche in some branch of science or scholarship.

As is made abundantly clear in his articles and personal letters, Dawson also learned the trick of gaining attention and respect by sounding authoritative. Treating material relating to a half-dozen different scientific disciplines, he employs a tone breezily assured and vaguely hortatory, managing to make what was actually a small amount of information hastily gathered seem as if it rode on a vast store of learning. Here he is holding forth to Woodward in the field of mineralogy:

> It does not so much effect the issue as the *cause* of columnar structure. Your mineral people would probably know? It is common to all colloids but starch is a hardy material to investigate. In a number of cases these prismatic or sub-prismatic cleavages have been mistaken for "flaking" of the ordinary form and the conchoidal flakes sometimes mask the cleavages or fissures.

Here he is disagreeing with Sir Arthur Keith—actually presuming to instruct the scientist—in the field of anatomy. Keith had pointed out that in the anthropoids certain teeth erupted at the same time:

> I reminded him that supposing this was a rule with the apes, we were dealing with an intermediate form and as the times for appearance of the 3rd molars in the human jaw were most variable and quite unconnected with the growth of the canines—that his new theory was not a very safe one. I also reminded him that the 3rd molars have already shown rudimentary characters both in the human jaws as well as certain of the anthropoids. The 3rd molar is often placed at such an angle as to look inward and is almost useless for mastication and the fangs are often not fully developed . . .

Here he is in the field of heraldry, in a short note to one of his Lewes neighbors. The note is signed "in haste," showing or implying that its contents came straight from his own personal knowledge on the topic:

> The unicorn became one of the supporters of the Royal Arms circa 24 March 1603 when the Sovereign union of Scotland with England

took place & was first used by James I of England (James VI of Scotland). The unicorn was a fabulous beast whose horn was suggested by the Narwhale (or rather it is a modification of a front tooth of that crittur). The unicorn supplanted the dragon or gorgon in Q. Eliz. arms on the right side of the shield. The city of London seems to have adopted the coat of Wyburn. Henry VII, VIII, & Edward VI had the greyhound, but all had the lion on the left side of the shield.

Here he is descanting learnedly on Wealden fossil plants, a topic he treats or refers to nowhere else. The occasion was a meeting of the Geological Society in London, just a month before Piltdown was announced to the world:

> Mr. Dawson observed as to the fossil plants now described, that it was rather remarkable to find side by side on one piece of Wealden rock remains of fossil ferns, *Matonidium* and *Hausmannia*, genera which have long ceased to exist in the Northern Hemisphere, but are very closely allied to the modern *Matonia* and *dipteris* respectively, which now grow side by side on Mount Ophir in the Malay Peninsula.

Here he is on a subject that must have been very near his own heart, forged or counterfeit paintings. The passage occurs in a letter to W. V. Crake of the Hastings Museum. The tone of easy familiarity with the offbeat topic is curiously revealing, or at least peculiar:

> Have you heard of a man at Hastings who is going about selling spurious works by "Old Masters"? I should be interested to hear as I have some reason to think that there is such a person—assisted by a young woman who offers them. He was formerly at Watling whence he issued all sorts of "speculative" pictures. I believe some of the trade are offering them. They are rather clever & the frames are not of the period and they are usually covered with glass. Like an old canvas—sometimes baked. You may warn your friends. If the glass is removed a pin in the whites etc. will reveal the paint soft!

By accident, as it happens, the one type of practical experience Dawson needed for his Piltdown effort—experience he did not possess and could neither easily acquire nor adequately imagine—handily came his way in 1907. Near the town of Eastbourne on the south coast, a human skull with a few bones was unearthed in the course of digging on a farm. Informed of the find, the Sussex Archaeological Society sent a member, John Ray, to investigate, and he invited Dawson to accompany him, probably as photographer. As the two dug around the skull and down through an extensive flint bed, a full skeleton was uncovered, that of a young man. Later that same day still another complete skeleton was turned up, this one of a young woman. Hours of painstaking examination, both of the bones and the soil conditions, indicated that the two skeletons were very ancient, probably going back to before the time of the Romans. From all indications, it was soon after his Eastbourne experience that Dawson decided to create Piltdown Man—certainly less than a year. By then he had learned and digested some valuable lessons about the natural disposition and appearance of human remains that had lain for many centuries in the layered soil of his native county.

It was in this same year that Dawson told Arthur Woodward about his encounter with a sea serpent, by any measure his weirdest, least understandable imposture. The story—which has him observing the creature for a full four minutes through binoculars—was presented quite seriously in a letter to Woodward of October 7, 1907, no less than eighteen months *after* the supposed sighting. "On Good Friday (1906) I was crossing the Channel from New Haven to Dieppe," he begins. The day was calm, the only wind coming from the movement of the steamer, the S. S. *Manche*. Between 2:00 and 3:00 P.M., thinking that they were in sight of land, Dawson went on deck. With his field glasses he was scanning the horizon,

> . . . when I caught sight of what seemed like a large, cable-like object struggling about. It was some two miles away right ahead of the ship. While trying to focus on the object as sharply as possible I heard two men (passengers) talking about it. One said, "Hallo! what's that coming, the sea-serpent or what is it?" The object had then shifted its course and instead of coming at us had turned on a side track of about 45 degrees from our port side, but offered a more extended and less complicated view.
>
> I could not see any head or tail but a series of very rounded arched loops like the most conventional old sea-serpent you could imagine, and the progressive motion was very smart and serpentine for when one loop was up its neighbor was down. I could see no detail except the long black arched line, dipping into the water at either end . . .

The loops were fully 8 feet high out of the water, and the length 60 to 70 feet at the smallest computation . . . I watched it receding from the vessel in an oblique direction until it entered the path of the sun's rays upon the water until I finally lost it.

Luckily, as Dawson says, he had his camera with him, a small Kodak, and he was able to take several shots. Unfortunately, and of course inevitably, on later developing the film (he was a competent photographer and did his own developing) there was nothing to see: "No detail of the sea or sky appears beyond a few yards." Afterward, he says, he talked about the sighting with his fellow passengers, exchanging cards with several. All agreed that the object had been "too bold and too large to be anything like a school of porpoises sporting after one another."

No response from Woodward to the sea-serpent letter is known. Nothing more was ever heard of the serpent. About the only pertinent observation to be made relates to the wide divergence in dates between letter and sighting. As with so many of his frauds, here again Dawson builds into his little scheme—even with *this* piece of clownish ephemera!—a lengthy time lag. If Woodward had shown any interest, of course, it would suddenly have proved very difficult to locate those other witnesses whose cards he had obtained, even if the cards themselves could be found.

The mind of a lifelong dissembler—what makes it incline now to this elaborate deception, now to that smaller one—is mostly past fathoming. But in the case of Dawson's sea-serpent letter at least some part of the motivation may be glimpsed. Arrestingly, it betrays the forger's utter confidence in being able to manipulate, almost to play with, the susceptibilities of his old friend, the country's leading expert in *marine* paleontology. The letter marks, perhaps, that moment in Dawson's life when he came to an overriding belief, no doubt to him quite thrilling, in his unlimited power to deceive.

The Harry Morris flint cabinet with its tantalizing series of handwritten notes about Dawson's "farce" still continues as a neat little mystery on its own. Since Weiner's initial efforts forty years ago, research has turned up nothing to throw further light on the Dawson-Morris link, so dealing with its implications becomes a matter of interpreting the little that is known. In this way, by sheer weight of necessity, some fairly definite conclusions present themselves almost unsought. It is clear, for example, that Morris did not suspect wholesale fraud at Piltdown and never imagined the true scope of the imposture. It is equally clear that he was the ultimate source of those persistent rumors of fraud emanating from the Lewes area, and which finally reached the American, William Gregory, to be preserved in his article of 1914.

The focus of the cryptic notes retrieved from the Morris cabinet is entirely on the subject of flint implements, study of which made up nearly Morris' whole interest in life. In the notes no allusion is found to the Piltdown bones, neither jaw nor cranium, none to the animal remains. While the canine does receive pointed mention ("there is every reason to suppose that the 'canine tooth' found *at P. was imported from France*"), this is nothing but a veiled and gossipy reflection of the earliest suspicions aroused by a foreigner, the French priest Teilhard. Read correctly, the phrase "*imported from France*" is meant to be figurative, an indirect way of accusing the priest of faking the canine.

Morris' concerns were all centered on those flints, and for good reason. With one of them, received directly from Dawson himself, he had been alert enough to test for artificial staining. His purpose in taking this action goes unstated, but it certainly was prompted by his unrivaled knowledge of Wealden flint beds: he simply didn't believe that the soil at Barkham Manor could yield such specimens. In fact his first thought was that "if they were genuine artifacts, they were Neolithic, and no earlier."

Coupled with the general distrust and dislike of Dawson because of the Castle Lodge affair, the staining of one flint was enough to put Morris on his guard (though, as he knew, the staining could have been done for other reasons than fraud). *Something*, he obviously felt, was wrong at Piltdown, something about the paleoliths and eoliths definitely, and who could tell how much else besides. Beginning in the summer of 1913, Morris talked freely with various confidants in Lewes in this questioning, accusatory strain. In the process whispers of fraud were generated that drifted from man to man in Sussex, at length filtering through to London and Gregory.

Among those who listened in fascination to the Morris complaints were the impressionable Guy St. Barbe and his friend, the equally naive Major Marriott. Because of what they heard from Morris, and then subsequently imagined, both "saw" much more during their unexpected visits to Dawson's Uckfield office than was actually present to the eye. The thought that the Piltdown mastermind would have been caught plying his secret trade in his own busy office, in a bustling public building on Uckfield's High Street, with the door unlocked, merits only a disbelieving smile. Whatever he was doing with those chemicals whose smell St. Barbe remembered, he was not manufacturing fakes (perhaps testing preservatives, for which there is some evidence).

The earnest if plodding Harry Morris, dedicated student of prehistoric tool-making, in reality had been accorded a signal honor by his friend Charles Dawson. As with those other intended victims, Conan Doyle and Lewis Abbott, he had been chosen to fill a prominent role at Piltdown, lending it further believability. Also as with those two, and more because of his own hardheaded independence where flints were concerned, he had managed to avoid the trap. Fortune, as it turned out, at the last smiled benignly on Dawson. But Harry Morris had come very near to being his one fatal miscalculation.

What it was that saved Conan Doyle and Abbott from becoming captives of Dawson and Piltdown, in the fashion of Teilhard and Woodward, is not on record. With Doyle, perhaps it was simply a heavy burden of work, both his own writing and his constant public involvement. Even so, he seems to have had a very close call, for Dawson went to some lengths to accomplish his bold purpose. For one thing, the flint arrowhead he picked up during his visit to the Doyle estate in the fall of 1911—"embedded and in view of us all"—and which he generously handed over to the delighted Sir Arthur, certainly came not from the rocky soil but from his own pocket.

12

Return to the Pit

On the second day of June 1912, a Saturday, a small party consisting of Charles Dawson, Arthur Woodward, and the young cleric Teilhard de Chardin arrived by automobile at Barkham Manor. They were armed with a variety of tools, including geological hammers and an assortment of brushes and sieves. On hand to greet them at the open pit with a supply of heavy-duty picks and shovels was the workman Venus Hargreaves. It was an exciting moment, the start of the Piltdown excavations—though for one of the three the excitement had a special quality of anticipation unknown to the others.

For this day, Dawson had spent no less than three years in secret and meticulous preparation. Carefully he had studied all previous fossil frauds and hoaxes—there had been several—as well as the legitimate fossil discoveries, in particular that of Heidelberg Man. Every fake fragment of cranium and jaw, every animal fossil, every flint, had been completed. Every intended move, to its last fine detail, had been planned. What makes it possible now to follow those hidden moves, step by step, is the plain fact of known fraud: *everything* about the excavations was designed to deceive. With that conceded, and knowing much of what actually took place at the pit, it becomes a fairly simple matter to enter Dawson's mind. Patient analysis of the known circumstances permits a reconstruction of his methods which stands as nearly certain, allowing here and there for slight differences in detail.

The party having gotten a late start—Woodward that morning had come down from London, not arriving until nearly midday—it was late afternoon before excavation procedures were settled and the digging begun. Working deliberately with his shovel, Hargreaves broke up the packed earth of the pit, heaping the dirt along the grassy margins to await sieving, or spreading it thin for a first, cursory inspection. At the same time the other three took turns either passing the loose soil through close-meshed screens or combing among the spoil heaps that soon ringed the pit.

In both operations every likely looking lump, large and small, had to be picked up, pinched in the fingers, peered at closely. A tap of a small hammer or pick now and then loosened some rock-hard accumulation of soil at rest on the ground or in a screen.

For this first day, Dawson's private plans were modest. They called for the discovery of only two small specimens, both of which at that moment reposed in his pocket (or pockets, since they would have been carried separate from each other). One was a part of the human cranium, from the occiput. The other was a fragment of a mammalian tooth, *Stegodon*. Both were meant to be found not *in situ*, but mixed in the detritus of different spoil heaps flattened on level ground some distance from the pit's edge. Probably Dawson's intention was to have the mammal tooth found first, by the sharp-eyed Teilhard, as a sort of preliminary. Then he would allow Woodward the thrill of discovering the cranial fragment, and the excavations would be off to a flying start.

But that hope, as it proved, went slightly wrong. Soon enough Dawson learned that the task of controlling the sequence and timing of the finds would prove rather a delicate one. His eager companions, despite their wide experience, often failed to recognize a planted specimen at their fingertips.

At first Teilhard overlooked the mammalian tooth lying where Dawson had surreptitiously placed it on the darkened soil of a spoil heap. At length, deciding he could wait no longer to plant the second piece, at a strategic moment Dawson positioned the skull fragment in another spoil heap, one awaiting inspection by Woodward. But Woodward, too, failed to pick out the diminutive bit of broken bone.

At last, as the day's work was nearing its end, Dawson reluctantly gave in and "found" the cranial fragment himself. (Permitting the bit of bone to lie exposed for days, or even overnight, posed too great a chance of its being lost for good.) Now occurred that unusually joyous reaction from Woodward, the display of youthful enthusiasm reported by Teilhard: "all the fire" hidden under his customary reserve suddenly flared in an expression of delight. A few minutes later Teilhard, his attention perhaps sharpened by the excitement over the first discovery, announced his own find of the *Stegodon* molar.

Aside from the jaw and bone implement, the actual planting of the specimens would have been the least difficult part. Depending on the physical location of the other three at any one time in and around the pit, Dawson could merely place the pieces on the ground by hand, while seeming to be conducting his own search. If he felt an open maneuver at any point might be too risky, he could drop the piece down the inside of a trouser leg, adroit placement of a foot conveniently depositing the specimen wherever desired (for this he would have prepared a special arrangement of his trouser pockets). What he could not predict or manage was the ability of others to maintain concentration, to see what they were supposed to see.

Actually embedding these fragments, as if in undisturbed soil, would have involved far too great a gamble, and in fact was not needed. An *in situ* condition called for much time and careful effort to simulate convincingly, and then there might still be some small detail to give it away to experienced eyes. Of all the finds at Piltdown, only the jaw, and to an extent the bone implement, required the framework of an *in situ* discovery.

In that first season at Piltdown a total of five cranial pieces were found, aside from the jaw, on five different days, all in spoil heaps. Only one of these was turned up by Woodward. The other four were spotted by Dawson himself, though it is certain that he would have preferred Woodward—or Teilhard when he was there—to be the discoverer of the larger share, and perhaps all. No doubt he made such an effort, only to find himself repeatedly frustrated when his target failed to discern the planted piece. Even the single fragment of skull that Woodward did find, on a weekend in July, came perilously near being overlooked. Picking the piece up from the spoil heap, as Woodward recalled, he thought at first it was only a bit of ironstone, like a thousand similar bits he had already handled. It was only, as he said, "after much inspection"—during which the anxious Dawson no doubt eyed him warily—and as he was about to discard it that he suddenly recognized his prize.

The three other cranial pieces that were "found" by Dawson were not brought to light until late in August, during a three-day weekend, one piece each day. Certainly this clumsy, last-minute bunching of finds as the season drew to a close was not Dawson's preferred method, but resulted from his inability to get the three pieces found earlier by others. Of course for most of the summer he had only Woodward to work with, Teilhard being available that year on no more than three or four occasions before he left for France. Having planted a fragment meant for Woodward's probing eye, when it went unnoticed, Dawson had either to "find" it himself or quietly retrieve it for planting another day.

If Dawson and Woodward dug at Piltdown on every weekend in the summer of 1912, into September, they would have been busily engaged there on some sixteen weekends, a total of more than thirty days, perhaps two hundred hours of digging. Indications are, however, that they missed several weekends or were on hand for only a Saturday or a Sunday, and once Woodward was there alone for a day at midweek. This meant that there were a great many hours of tedious sifting, digging, and hauling in which nothing much happened. But while the hominid finds were few in this period, Dawson arranged for a satisfying number of mammalian discoveries to fill the time. Most were teeth from a half-dozen different animals, all yielded by spoil heaps. These items, along with the flints, both paleoliths and eoliths, were of value for dating purposes, but they also continually buoyed hopes for added hominid discoveries.

Here again, though, Dawson had considerable trouble in leading Woodward, and Teilhard when he was there, to spot his planted goods. Woodward, unaided, picked up only a single one of the total of eight animal teeth found, Teilhard none after that first day. Dawson himself had to "find" the other six. With the three paleoliths (worked flints) he had better luck, Woodward and Teilhard each managing to come up with one. The so-called eoliths were somewhat harder to recognize, and the result was that Dawson had to produce all nine.

The most severe challenge faced by the forger in that first season, of course, was the jawbone. This was much too large an object, too bulkily obvious, to allow it to be found lying on the surface in a spoil heap. Even to have reached a spoil heap it would supposedly have come out of the ground on Hargreaves' shovel, and in that case the workman himself, or someone nearby, could have been expected to notice it rather quickly. The possibility of damage by a probing shovel had also to be avoided. For those reasons, and to achieve its maximum impact of authenticity, the supremely important jawbone had to be found *in situ* in the pit, or to appear as if it had.

Further, since it would be virtually impossible to simulate an *in situ* condition in the hard-packed earth of the pit bottom, Dawson himself must make the discovery. At the same time Woodward must be present to witness it, though being deprived of any chance to study the spot, or even to be concerned whether the jaw's earthy bed had previously lain untouched. It was a knotty problem, indeed, the fraud's real crux. Typically, Dawson's solution to it was an ingenious one, actually his boldest, and his rashest, decision of the entire campaign.

Since there must not be the least risk of anything going wrong, burial of the jaw could not be done much beforehand, by night or day. Leaving it at rest too long in the Barkham Manor gravels would be foolhardy. Dawson would have to insert it in the soil of the pit bottom on the same day it was "found," and only shortly before. He would have to perform the operation while the others were present and moving freely around the site. This was feasible since he could await the proper moment to act, when his companions were busy elsewhere and he himself was crouched down in the trench pretending to search. It need not be done on any particular day or weekend. No move need be made until all the conditions were favorable. As it was, Dawson bided his time only until the fourth weekend of June.

The last necessary element of the mechanism for the jaw's unearthing concerned *how* it was to be accomplished: just when, and in what exact manner, it would be freed from the enveloping soil. Here, it must be conceded, Dawson was at his best, for the method he chose, while fittingly dramatic and effective, also posed by far the greatest gamble. All would depend on precise timing, patience, and a steady hand. At last, on Saturday, June 23, while he was alone in the pit, with Woodward occupied nearby, he inserted the jaw fragment a few inches into the blackened soil. Moments passed, and then came his chance.

"I struck part of the lowest stratum of the gravel with my pick," Dawson later explained, "and out flew a portion of the lower jaw." Just at the moment he wielded the little geological pick—the evidence shows this clearly—Woodward had come up. At the critical instant he was actually peering in Dawson's direction, in fact was looking at the very spot where the blow fell. The timing, force, and aim of the decisive stroke were all perfection, for when the jaw made its sudden appearance out of the earth, Woodward was utterly persuaded that he had seen the dark L-shaped bone being dislodged, as he ever afterward claimed, from "untouched remnants of the original gravel." Untouched the gravel of the pit floor certainly was, all but the few square inches in which for a matter of minutes the jawbone had nestled.

So well planned, so deftly executed was the crucial maneuver that Woodward instantly became the principal and most important witness to the jaw's discovery. With his own eyes, he proudly stated, he had seen it "fly out in front of the pick-shaped end of the hammer." Never did he think to ask why the experienced Dawson would have *blindly* delivered a blow hard enough to make the jaw actually jump into view—"fly out"—when he should have been gently probing the soil. In that particular circumstance there was no condition that called for a forceful blow, not to dislodge a visible or suspected artifact, certainly not to probe new ground.

About inspecting the gravel for signs of prior disturbance Woodward says nothing, and that is certainly because Dawson promptly obliterated the evidence. While Woodward's fascinated eyes followed the jawbone skittering along the pit bottom, Dawson in feigned excitement was no doubt trampling heavily on the earth all around the immediate area.

In addition to careful planning and masterful execution, there was one added factor Dawson knew he could count on to seal and ratify his trickery with the jaw. He knew that his longtime friend and coworker trusted him. He knew that Woodward would never think to question his claim, implied or spoken, that the pit bottom had been undisturbed. Perhaps the most damning thing that can be said of Dawson, in a human way, is that in all he did as a forger he never hesitated to trade on the fact that his friends and colleagues liked and trusted him. After the exposure there were many who in recalling the affable, obliging solicitor, could not connect him with an act of betrayal.

The famous Heidelberg jaw, so crucial in the evolutionary picture, was discovered in Germany in the fall of 1907. The following spring it received its first public announcement, becoming the talk of scientific circles. In England newspapers and more serious journals gave it extended coverage, all of which was certainly read by Dawson. Heidelberg's role as the igniting spark of Piltdown, suggested long ago, appears well established: it was a desire to repeat and surpass that huge success, his eye still on a coveted Royal Society fellowship, that propelled Dawson to action. From that fact is traced the first undoubted link with Piltdown, for it was in 1908, probably the fall, that Dawson made his opening move in the fraud when he told his friend, the chemist Samuel Woodhead, about finding a piece of the skull. Soon afterward the two spent a day together at Barkham Manor in a fruitless search for more. This early link with a competent witness was pointedly stressed by Dawson in all his published references to Piltdown.

The unsuspecting Woodhead became, in other words, the forger's initial victim, representing what was from the start an integral part of the daring scheme. Corroborating witnesses, men whose word, character, and competence would be seen as unassailable, were to be involved at every turn. In different ways and at different times Dawson targeted at least half a dozen of his friends and acquaintances, all men of high repute. With most he failed, even losing Woodhead eventually. But one, Teilhard de Chardin, more than made up for all the others. Here was an ordained priest, young, idealistic, naive, a keen fossil hunter who was also a qualified scientist—the perfect dupe.

When in 1913 it came time for the important canine to be discovered, no better medium for its delivery could be imagined than the priest. With Teilhard, in fact, if Dawson had had his way, there would have been a direct link with Piltdown, similar to that of Woodhead, even before the start of formal excavations. In April 1912—by then the two had known each other for three years—Dawson visited Teilhard at the seminary. With him he brought some pieces of what Teilhard in a letter of the time described as "a very thick" human skull, accompanied by three fossil animal teeth. These fragments were shown him, wrote Teilhard in a letter to his parents, "in order to stir me up to some similar expeditions, but I hardly have time for that anymore." Soon, though, the relentless Dawson had his way. Only five weeks later the priest was on hand at the pit, eager for the first day of digging.

Teilhard's unsuspecting involvement with the canine in 1913 affords a classic instance of Dawson's peculiar genius for well-designed chicanery. That the canine, as envisioned by Woodward in his reconstructed model, would become a center of controversy was of course a development that Dawson could not have foreseen. But no sooner did it arise than he grasped the renewed opportunity to further strengthen Piltdown's claim. By early summer at the latest he had succeeded in fabricating the disputed tooth and was awaiting the opening of the season's work at the pit.

The only available candidate he had at this time for the role of the canine's finder was Woodward, young Teilhard having returned to France the previous year. But for some reason Dawson delayed the canine's unveiling for weeks, and then unexpectedly in July he heard directly from Teilhard. His superiors in France, it developed, were sending him back to England for a month's stay at the seminary. Unhesitatingly Dawson formed his plan. By mail he extended to Teilhard an invitation to stay as his guest at Castle Lodge on the weekend of his arrival, which Teilhard gladly accepted. None of this is conjecture: in a letter of 1954 to Weiner, Teilhard states, "In 1913, my staying overnight in Lewes (and the trip to Piltdown) was *prearranged*." The italics are his.

Especially compelling in the context of the forgery is the incident of the canine, for by means of it Dawson's manipulations may be followed more closely than is possible with most of the other finds. Fleeting but revealing details, inadvertently preserved by the participants, allow for a satisfying reconstruction of how it was done.

The priest's schedule of duties at the seminary would have been Dawson's initial concern: did it permit his attendance at the pit on every weekend? The answer was a bit disappointing: Teilhard would be free only for the weekends of August 9–10 and 30–31. In between, he was to take part in an extended spiritual retreat at the seminary, the main reason for his return. Then on September 2 he would depart Hastings for the trip back to France. Dawson thus could be sure of his prey for only two weekends, four days, even less should any interference arise. On all four of those days at the pit he would be carrying the tooth with him, ready to act.

There was, however, a slight complication. For the weekend of the ninth Dawson had already arranged to implicate Woodward's wife, Maude, who would be drawn into the finding of the nose bones. This occasion could not be readily switched, for Maude Woodward's visits to Barkham Manor had been very few. Probably it had taken a good deal of cajolerie on Dawson's part to persuade her to make even this one August visit. Whether he also set his trap for the innocent Teilhard on that first weekend escapes detection. Very probably he did, not wanting to chance letting matters go too late, only to have the expected little drama misfire when Teilhard failed to spot the darkened tooth lying in the rubble.

Only one weekend remained, which might well be reduced to a single day if other business should occupy Teilhard on what would be his final Sunday in England. Dawson's last reliable opportunity would occur on Saturday, August 30. To help ensure Teilhard's presence that day, he told him it would be the season's concluding day of excavation (it wasn't, but the departed Teilhard never knew that).

The digger Hargreaves did not show up at the pit for the weekend of the thirtieth, nor was there a replacement for him, certainly a deliberate move on Dawson's part. Without Hargreaves steadily filling buckets and pails with earth to be sieved, attention could be focused on a more relaxed and leisurely inspection of the spoil heaps already spread along the grass—and it was in a spoil heap that the canine was to be planted. Further, Dawson now adopted a new search technique, one that made good sense but of which there is no sign of earlier use. He had the spreading yards of spoil divided into small sections and visibly marked off, perhaps by borders of string—"mapped out in squares," as he said. With this grid system (even then a fairly standard procedure in archeological digs) the searcher's attention would be more powerfully concentrated on one square at a time, each square being canceled when inspection was completed. With the searcher crouched down on his hands and knees, "crawling," as Woodward described it, no inch of the spoil would be missed.

Fortunately the exact moment of the discovery was precisely recalled by Woodward soon after. Teilhard, he remembered, had been digging in the pit for some time, had then been relieved of shovel work and set to looking over the squared-off heaps. In his description Woodward continually writes "we," meaning himself and Dawson, but in this instance, putting the priest to work on the spoil heaps, the lead was surely taken by Dawson, of course in slyly casual fashion.

Some minutes passed in silence, all three men sweating profusely in the sun as they were preoccupied with their individual tasks. Suddenly Teilhard excitedly called out—"exclaimed," he said later—that he had found a tooth! The other two, however, were slow to react ("We were incredulous"). Dawson even called out dismissively that Teilhard had probably been fooled by a bit of the ironstone that littered the area, and "which looked like teeth." When Teilhard insisted, the two at last climbed reluctantly out of the pit and walked toward their grinning companion.

"There could be no doubt about it," wrote Woodward, later recalling the moment. It was indeed the veritable canine. Greatly exhilarated, all three promptly dropped to their knees at different squares and began eagerly searching: they "spent the rest of the day until dusk crawling over the gravel in a vain quest for more."

Ten days later to his parents, Teilhard briefly recounts his "lucky" find, stressing how "very exciting" was the experience. Some of this original exhilaration still remained with him forty years later, it seems, helping to cloud his judgment over the idea of forgery. When Weiner wrote him in 1953 to explain about the exposure of the fraud, Teilhard responded strangely. The canine, he wrote, had been "so inconspicuous amidst the gravels" that to him it seemed "quite unlikely that the tooth could have been planted." Despite having just been officially informed that the tooth was in fact a fake, and therefore could have occurred at Piltdown only by planting, this world-famed scientist shows himself still deeply reluctant to admit that in his youth he could have been so badly fooled. His muddled and illogical reply to Weiner uncovers a proud spirit, one unaccustomed to deceit, and all too human. But very soon, in a cooler moment, his old strong suspicions about Dawson must have come flooding back.

Why Dawson would have deliberately gone on from his success with the jaw and the canine to undertake the really daring gamble of the bone implement is a question that requires but has never found an answer. However, that answer, wholly adequate, may be readily extracted from Woodward's own few scattered remarks. For Woodward the implement was "unlike anything hitherto known among the handiwork of prehistoric or primitive modern man," affording the best proof of Piltdown's antiquity. The mammalian remains from Piltdown, the animal teeth, he had come to feel after his first excitement cooled, were "insufficient to date *Eoanthropus* with exactness," leaving only the problematic witness of the gravel. For Dawson, that small slippage in Woodward's certitude gave its timely warning, and he moved rapidly to produce the implement. The gamble, a real one requiring skill, confidence, and another hurried search in paleontological literature, paid off in wonderful fashion when Woodward accepted the result as "probably the only specimen of real importance" when it came to dating the Piltdown finds.

Concerning the implement's unearthing, a single circumstance is sufficient to indicate how it was all managed: the generally overlooked fact that it came from under a hedge. In his formal presentation of the implement to the Geological Society in December 1914, Woodward states that it was found "about a foot below the surface in dark vegetable soil, beneath the hedge which bounds the gravel pit." His personal account of the incident, written less than two years later, adds an important detail. The find was made "beneath part of the hedge which Mr. Kenward had allowed us to remove."

Since the more than foot long, yellow-brown implement was deliberately planted, and since Dawson could not have managed such a burial while the others were present, it is clear that it was put into the ground some time before the late June afternoon on which it was found. Burial could have been a day before, it could have been weeks. Most likely is a few days, placing the act between two weekends' digging.

Dawson's access to the pit by this time, of course, was free and un-hampered. Spending almost every business day in his law office at Uck-field, an hour's walk or a little more from Barkham Manor, he must often have paid casual, apparently nonworking visits there. Even if he happened to be met at the pit by other visitors, or by a member of the resident Kenward family, the encounter would have had no particular significance and would have been quickly forgotten. It was on some such occasion, when he had the pit area to himself, that he planted his final fraudulent artifact.

The soil that yielded the bone, though lying under a hedge, had already been disturbed. A good portion of it, said Woodward in his official report, consisted of spoil from previous digging, and this, he thought, accounted for the presence of the implement: it had been "thrown there by the workmen with the other useless debris when they were digging" in the pit immediately adjacent. Aided by Woodward's at times astonishing naïveté—his curious willingness to believe that the workmen, who had been promised money for anything unusual, would have flung the bone away—Dawson made good use of his opportunity. The large implement, broken almost in two, could be hidden in disturbed soil, greatly simplifying the burial and reducing any later risk of detection. There was no need this time for a full-fledged *in situ* condition. The hedge itself would be proof that the slab of bone had lain there for a number of years, perhaps as much as a decade or more.

Getting the implement safely into the ground would also have been an easy matter, for it certainly was not put in from the surface, but from the side. The descending inner wall of the widened pit was less than two feet distant from the hedge. Dawson needed only to find the correct spot on the pit wall, a foot down from the top, and then burrow horizontally. The appearance of the wall's face after the insertion, any marks of disturbance, would pose no problem. Signs of interference would be taken as a normal result of the digging previously done. A final advantage of this horizontal burial, accomplished with Dawson concealed in the five-foot-deep pit, was the protection it afforded from accidentally prying eyes. Even if interrupted at his task he need offer no explanations.

Removal of the hedge at that particular spot can only have been Dawson's doing, and what reason he offered Woodward for the action would be interesting to know. In any case, Woodward by now was almost completely under Dawson's spell, so it would have taken little by way of explanation to satisfy him. Probably Dawson mentioned something about wanting to probe an extension of the deeper, flint-bearing gravel bed.

With the hedge gone, Hargreaves was assigned to break up the newly exposed surface. Nearby stood Woodward, watching. Several blows of the mattock had been delivered when suddenly "some small splinters of bone" flew up. Immediately Woodward halted the workman. Kneeling down he dug deep with his hands, felt something hard, and "pulled out a heavy blade of bone." For Dawson this was a lucky deviation from the plan, for he could not have anticipated Woodward's seeing the flying splinters or his prompt action with his bare hands. No doubt Hargreaves had been Dawson's intended victim, another innocent witness to be drawn into his net.

While washing the clinging dirt and clay off the bone, Woodward recalled, Dawson pointed out that one of its ends looked as if it had been broken straight across. Maybe the other piece was still in the ground? Bending down, he plunged his hands into the dirt and "soon pulled out the rest of the bone." The implement, of course, had not been broken in the ground but had been buried in two pieces, more of Dawson's minute care. When whole, it would certainly have been too cumbersome to carry concealed on his person. It would also have been too troublesome for convenient burial, especially in conditions calling for haste and a minimum of effort.

With that, the series of "discoveries" made during three years in connection with Piltdown Man came to an end. The other supposed finds, those made at Sheffield Park (Piltdown Two) and Barcombe Mills, require no review. They were never in the ground anywhere. After being fabricated in Dawson's private workshop in the cellar of Castle Lodge, they went straight to the hands of Arthur Woodward.

Concerning the Barcombe Mills material, however, its *purpose*, the reason it existed at all, there is still something to be said. Reported to Woodward by Dawson as early as July 1913, its content precisely matched that of the later Piltdown Two: a molar and two cranial fragments, one of which retained a bit of the orbital ridge (the added fragment of the cheek in the Barcombe Mills cache is not germane). This resemblance requires an explanation, of course, which can only be that Dawson's original schedule called for him to produce a second Piltdown Man much earlier than actually happened, a full two years earlier. If the plan had not been altered, Barcombe Mills and not Sheffield Park would have been designated as Piltdown Two. What interfered with the original plan? Judging by known fact, the answer is obvious: the sudden need to introduce into the developing picture a more dramatic element, the canine, found on August 30—or rather the *opportunity* to introduce it.

Piltdown Two was not, as has been regularly suggested, an afterthought, a protective response to the rising opposition of critics. From the start it was an integral part of Dawson's scheme, intended to add its weight of testimony within six months of the first announcement, a fact sharply attesting to his bold if distorted foresight. But when, unexpectedly, the dispute over the form of the canine, as reconstructed by Woodward, claimed the spotlight, matters quickly altered. Taking advantage of the new development, Dawson was able on the instant to change direction, setting Barcombe Mills aside temporarily in favor of the controversial tooth. As events proved, of course, he was exactly right in his choice. From the furor over the canine he gained far more in the way of public attention than would have resulted from Piltdown Two, as well as further corroboration.

Afterward, Dawson continued to withhold Piltdown Two, awaiting a propitious moment for its introduction. Not until 1915 was it brought into play—and then in the improved version of Sheffield Park—when he decided that still another sensation was needed to add support. Here, however, while he did manage to put to rest the fight over linkage of cranium and jaw, he lost his hoped-for sensation. Woodward, hesitant and undecided about the new finds, hung back until 1917. By then all that anyone really cared about was the World War.

If Charles Dawson was innocent of all complicity in the Piltdown affair, then it is plain that everything he told about it from his own knowledge, barring oversights, must be true. His story about the diggers at the pit finding an intact skull, which they took to be a coconut, smashing it to fragments while saving a piece for him, *must* be true if he was blameless. But the coconut story is the very element in the tangled tale of Piltdown's beginnings that, in strict logic, *cannot* be true. Once the fact of forgery has been established, especially a forgery so far-reaching and demanding such great pains to prepare, any such haphazard beginning is ruled out.

The notion that the original forged cranium was planted intact in the pit, its discovery left to the whim of busy, unconcerned diggers, is nonsense. The forger in that case would have lost control of his project at the start, for he could never have been sure of the outcome. The diggers, coming upon the cranium in the earth, might recognize it as a skull, or might not. They might take it up whole, or might smash and ignore it. They might keep it to sell to some local antiquarian (on hand in every village), or one of the men might decide to take it home as a keepsake for his mantelpiece.

Further, if the men did find and proceed to smash a skull, deliberately or otherwise, the breakage pattern certainly would not be the particular one required by the forger. Here the actual risk would be very large, for no one could tell how much and what sort of damage might be inflicted by men merrily flailing away with heavy picks and spades. In the end, with the skull perhaps reduced to a pile of splinters, the men might well forget to tell the forger anything at all about their find.

Even less reasonable is the idea, first mentioned by Weiner, of a substitute cranium being planted in the pit, which was afterward discarded in favor of the prepared fragments. Against this approach all the same objections apply as with the actual *Eoanthropus* skull, and there are a number of added disadvantages. Bits and pieces of the smashed substitute skull could remain in the soil to be inconveniently found later on. Also, if the men should take the substitute, whole, to some local expert, and publicity ensued, the forger would have effectively lost his all-important gravel bed. Unable to sow the same location twice, he would have to begin all over again at another site. But all such objections may be wrapped up in the one assertion that so ingenious and resourceful a planner as was Piltdown's creator would never have chanced incurring the least loss of control in the working out of his meticulous design.

The fuzzy nature in general of the coconut story can be readily demonstrated by a glance at a succinct version supplied by Arthur Woodward in his book, *The Earliest Englishman*. Within Woodward's brief text, which agrees with other tellings, can be found two blatant contradictions, either of which is enough to invalidate the original story. Dawson, writes Woodward, drawing on his many intimate talks with his friend, having come across the flint-bearing gravels at Barkham Manor,

> . . . asked the labourers to look out for bones or teeth or anything strange, saying that he would call again to collect their finds and give them a suitable reward . . . One day . . . the men dug up what they thought was a coconut, and they felt sure that this was the kind of thing which would please their curious and presumably generous friend . . .
>
> But as it was a little bulky to keep, they broke it with a shovel and threw away all but one piece which they put in a waistcoat pocket to show Mr. Dawson on the first opportunity. When he came round again the men produced their find and described to him the "coconut" from which they had broken it . . .

Glaringly obvious is the inanity of the claim that the men kept only one piece of what they found as of possible interest to Dawson while throwing the greater part away. If there was a chance for financial reward (diggers in British soil were all quite aware of this possibility, even without Dawson's offer), why throw away any portion of the object before they had Dawson's reaction? Why keep only a single piece? The men had only to drop the whole thing on a shelf in a shed awaiting their man's next visit. Once dwelt on, the second contradiction is even more blatant. The men, supposedly, broke up the skull because they thought it was only a worthless coconut. But did they really expect that a bit of coconut shell would be of interest to an antiquarian?

The coconut was merely another Dawson invention, meant to lend color and realism to the story of the smashed skull. But he used it only in conversation. Never did he put the word into print, no reference to a coconut occurring in his official or unofficial articles on Piltdown. Deliberately, he left that to others. (One of the important victims of the coconut story was Mabel Kenward, daughter of the Barkham Manor tenant. She lived into her nineties insisting on the reality of the smashed coconut, never conceding that the Piltdown skull wasn't genuine.)

Rigid, precise control of the operation at every stage, leaving as little as possible to chance, was the hallmark of the Piltdown fraud. Within those boundaries there was simply no room for the uncertainties of a planted whole cranium. No intact skull of any sort was recovered at any time from the pit. Of the hominid remains, only the five cranial fragments and the jaw ever touched the earth of Piltdown, and those fleetingly.

It was Dawson's habit of calculated daring that imposed on the Piltdown story the engaging idea that it began with an accidentally smashed skull. As he expected, that arresting picture subtly thwarted the initial tendency to doubt, shifting attention away from what is potentially the weakest part of any fraud, its start. Clearly he could have had no fear that anyone would try to investigate the claim, to trace and speak with the workmen—and no one did, the gathering of such information then not being deemed vital. If the attempt *had* been made, at every turn there would have arisen endless frustration and confusion over names, dates, and who had worked where, when. Here was a neat repetition of a pivotal aspect of the Beauport statuette fraud, in which Dawson actually identified a workman by name, though not until twenty-eight years after the fact.

A deliberately murky record of dating and chronology, darkened further by the passage of time between discovery and announcement, was Dawson's shield. Those obscuring elements, so innocent on the surface, he built carefully into a good many of his impostures, with Piltdown Man most cunningly.

Epilogue: Finished?

Sitting opposite me in an armchair at her home in Canterbury, dressed in a red velvet jogging suit, was the only person still living who had dug in the gravel pit at Piltdown, starting at age eleven, alongside Charles Dawson and the others. She was a dainty woman possessing an easy smile and quiet charm. If I had not earlier been given an introduction I should still have recognized her, for she was the image of a close-up photograph I had seen of her father, Sir Arthur Smith Woodward. Her hearing on the right side, she was explaining, was not good, but what could you expect at ninety-two?

The fascination of the deep open pit at Barkham Manor Margaret Hodgson remembers well, remembers also Charles Dawson with his balding head and friendly manner. In summers her father and mother, with her brother Cyril and herself, would spend weeks on the Sussex coast at Seaford, frequently traveling up to Piltdown for some digging. This went on not only during the early years of the discovery, but for long afterward.

My concern at the moment was primarily with her father, and his role at the center of the Piltdown affair. Curiously, accusations against him, as either forger or confederate, had always been surprisingly absent from the discussions, few even thinking such a thing possible.

When the fraud was first revealed, in the initial resentment and confusion immediate suspicion had been directed at him, but it soon lapsed. In answer to my question, his daughter said she could recall no particular disturbance in the family circle caused by such charges, public or private.

But, I inquired, during his five-year involvement with Dawson and Piltdown, and through the rest of his long life, did her father never suspect, no matter how slightly, that something, somewhere, might be wrong? Her answer was prompt and obviously sincere: she could recall no sign of any such wariness on her father's part, no word or look that would indicate the smallest doubts. There, I decided, I would leave the matter, persuaded that the woman had told only the truth about her revered father as she understood it.

Unfortunately, the available evidence tells a somewhat different story, and that story, too, undiluted, must be made part of the record. While the charge at this late date cannot be proved, it is morally certain that Arthur Woodward did in fact for long entertain nagging suspicions about Dawson's activities at Piltdown. It is a mere matter of granting Woodward a certain level of intelligence: no experienced professional scientist could have known all that he knew without feeling uneasy, without a gnawing fear that all was not as it appeared.

Most obvious in this regard was the strange muddle that clouded the origin of the Sheffield Park remains. These specimens—that were to prove so pivotal—reached Woodward in stages during the first half of 1915. But for no apparent reason he failed to make them public, holding them back in secret for more than a year and a half. Not until well after Dawson's death did he reveal their existence, and then in explaining their provenance he barely avoided telling an outright lie. Privately, as is known, Woodward admitted that "Dawson would not tell me the exact place" of the Piltdown Two discovery, and Dawson never did tell him.

Certainly this astonishing refusal was the main reason for Woodward's decision to withhold the new bones. He could hardly ask others in the scientific community to believe in their authenticity and supreme importance while he himself remained wholly ignorant as to their origin.

Whatever reason Dawson may have given for his peculiar and annoying silence, it clearly failed to convince Woodward, for otherwise he would not have hesitated over the new finds. This hesitation, at length if not at first, surely must have generated moments of deep puzzlement bordering on actual suspicion, inevitably to be bolstered by several earlier peculiarities. There was, for example, the Barcombe Mills find, which Dawson with such enthusiasm had reported as a possible "descendent of *Eoanthropus*." Could so shrewd a man as Woodward, a world authority on fossils, have compared the teeth and bones from Sheffield Park with those from Barcombe Mills, the two sets of remains so strikingly similar, without feeling a disturbing apprehension that *something* was amiss? At the back of his mind would there not have stirred other unsettling thoughts, for instance about the wonderfully fortuitous canine, so curiously and exactly matching his predicted model? Surely at times, especially after Dawson's early death, he must have pondered long about the story the solicitor told of Piltdown's beginnings, a story never more than imprecise and rambling.

Finally, in his own home in London in September 1913, it is nearly positive that Woodward heard an overt charge about the possibility of fraud at Piltdown, perhaps even joining in an animated discussion on the topic. It was on that evening that there occurred the social gathering, mentioned earlier, attended by a small group of paleontologists, including Teilhard, then on his way back to France. In that group was the American William Gregory, whose magazine article that fall put in print for the first time the rumor that Piltdown, as Gregory wrote, was a "deliberate hoax," and that all its bones had been "artificially fossilized." But whether in his own home or elsewhere, then or later, the same rumor would have found Woodward, as it found the transient Gregory.

In connection with the revelation of fraud in 1953, the investigating team was at some pains to protect Woodward's scientific reputation, and by implication assert his innocence, all quite understandable in the context of the times. To that end, the official report goes so far as to aver that the arguments Woodward put forward in 1912 and later "made a coherent and convincing case." The evidence available to Woodward at that time, explains the report, could not "be shown to be untenable." But that statement far exceeds the reality, a fact easily demonstrated. Here is a single instance out of several: if Oakley in 1953, using a magnifying glass, could so speedily find the telltale abrasions on a Piltdown molar, surely Woodward might have done the same in 1912.

In many of Woodward's anatomical interpretations on the cranium and jaw he made a deliberate choice between opposites, even where the possibilities were of apparently equal weight. Often he expressed himself as positive where the data were insufficient. Invariably his decisions went in favor of the whole Piltdown skull as a unified fossil of great antiquity. Though inwardly convinced that he was acting with high professional probity, in reality, as now can be seen, he was both hasty and biased. His country's role in world science, and his own fervent hopes for rising in that world, at the British Museum in particular, were the subtle factors that led him astray.

That conclusion, lamentable as it may appear, is fully justified by all that has been demonstrated in the foregoing pages. But more specific evidence of Woodward's special culpability is not lacking. It concerns the handling of Piltdown Two, the moment and the situation in which he made perhaps the most fateful decision of his life.

In February 1917, after a delay of a year and a half, Woodward finally presented the Sheffield Park bones to the Geological Society in London. In doing so, he blithely overlooked Dawson's strangely adamant refusal to tell just where he had found them. Instead of publicly admitting to ignorance on this point, as would have been proper, he placed himself and his reputation squarely and shamelessly in the breach. The site of the new discovery, he confidently announced, was a certain "large field" in the Sheffield Park district, a field that, he took care to add, he and Dawson had visited "many times." Hearing that downright claim, his listeners could only have assumed that he knew the exact spot of the find. Most people would also have quite naturally concluded that Woodward himself had been present at the discovery, at least had been directly involved.

The decisive impact of Piltdown Two within the scientific community was in great degree a result of the prevalent belief that Woodward had been intimately concerned in it from the start. If he had rejected the Sheffield Park bones, or had reserved judgment on them, if he had been more forthright and honest about their uncertain background, Piltdown Man would soon have lost its place in the evolutionary argument. Persistent doubts and objections as to the linkage of jaw and cranium, growing from the outset, would eventually have carried the day.

Not then, not ever afterward, did Woodward further identify or even refer to the "field" he specified so prominently in his official report. In his book of later years, *The Earliest Englishman*, he minutely rehearses everything to do with the creature, his "everyday life," his physical appearance, the tools he made, the animals he lived with and hunted, his place in man's evolutionary progress. But the supposedly precious remains of Piltdown Two have almost vanished from the scene. In the book's first chapter, promisingly entitled "The Story of the Discovery," they naturally deserved to be highlighted. Instead, they receive no mention whatever. Only on a later page are they referred to, and then in a total of four lines, and only in passing. The "field" of their discovery has evaporated, to be replaced by a "patch of gravel," not otherwise located or described, but situated vaguely "about two miles" from Barkham Manor. Their significance, the decisive role they played in gaining world acceptance for the fossil, is ignored.

Perhaps the most pathetic aspect of the whole Piltdown story revolves on the picture of an aging Woodward returning to the pit to dig every summer except one during twenty or more years after the untimely death of Charles Dawson. Many hundreds of additional hours were spent combing through the old gravel as he slowly widened his search beyond the original site. But his interest was no longer solely scientific. In large part, it had become a dogged hunt for personal reassurance, conducted in the wan hope that he might uncover final, absolute proof of Piltdown's reality.

Sir Arthur Smith Woodward, sad to relate, did not die a wholly contented man.

Notes and Sources
Selected Bibliography
Index

Notes and Sources

The left-hand margins display *page* numbers. For quoted matter the first few words of the quotation are repeated, enough to make identification sure. Citations are given in shortened form and may be identified by a glance at the Bibliography. Mingled with these citations is some added information and comment that may be of interest or value.

The following abbreviations are used:

QJGS—Quarterly Journal of the Geological Society of London
Papers—The Piltdown Papers 1908–1955: Correspondence and
Other Documents Relating to the Piltdown Forgery, compiled
with notes by Frank Spencer, Oxford University Press, 1990.
SAC—Sussex Archaeological Collections
SAS—Sussex Archaeological Society

PROLOGUE:
Unfinished

xvi "The most troubled chapter"—Weiner, *Forgery,* 204.
xvii My decision to omit from this treatment of the Piltdown story some of its more technical and peripheral aspects (for instance, Edmunds on the age of the gravels, Shattock on skull thickness, Marston on the jaw) is a considered one. None of these matters played any essential part in the outcome, and to retain them, once the fact of fraud has been established,

results in unnecessary complexity. Such material is of real interest only to professionals, and then in a quite specialized way.

xviii The Piltdown bones: I am grateful to Dr. Christopher Stringer of the Natural History Museum, London, for arranging my inspection of these at the museum in September 1993.

xix Hrdlicka's visit to the museum: Hrdlicka, "Early Man," 79–80.

xx "the degree to which"—Tobias, "Appraisal," 243.

CHAPTER ONE:
Curtain Falling

3 There is no formal biography of Arthur Woodward. I have used the following: 1) manuscript autobiography, unpublished, at the Natural History Museum; 2) obituary, London *Times*, September 4, 1944; 3) *Geological Magazine*, January 1915; 4) *Dictionary of National Biography* (1941–50), 974–5; 5) *Obituaries of the Fellows of the Royal Society*.

4 "The most important thing"—*London Evening News*, March 3, 1924. The interview has a comment by Woodward that shows his aloof temperament while at the museum: "We receive all sorts of silly inquiries from cranks, such as what do you think about the Deluge? [Such letters] are disregarded." Questions about the Flood, of course, were hardly silly at a time when the historicity of the Bible was a much-controverted question.

4 "jovial"—Mrs. Hodgson in conversation, October 1993.

4 "I was deeply"—Ms. Autobiography, Natural History Museum. Same for the next three quotations.

7 "a long past world"—*Nature*, July 30, 1938, 217.

8 "the pictorial art"—Dawson & Woodward, *QJGS*, V. 70 (1914), 100.

8 "The bone described"—Sollas, *Hunters*, 529.

8 Woodward on the Sherborne horse: *Nature*, 117 (1926), 86. For more recent discussion see *Antiquity*, 53 (1979), 211–16; *Nature*, 283 (1980), 719–20; *Antiquity*, 55 (1981), 41–6; Blinderman, *Inquest*, 186–8; Spencer, *Forgery*, 170–1.

9 "expressing himself at"—Woodward, *Earliest*, 53.

9 "He certainly was a man"—Ibid., 55.

9 "indeed a man of the"—Ibid., 74.

9 In the 1930s and after, the questioning of Piltdown's true worth and value as a fossil (but with no thought of fraud) was quite general. The attitude is handily summed up by a brief passage in a 1931 book review by Grafton Elliot Smith: "The widespread suspicion of the authenticity of the Piltdown man as a valid genus is notorious, and the chief reason for the lack of agreement in human palaeontology. Even today many Continental anthropologists refuse even to refer to it in treatises on fossil man or, when they do so, brush it aside as being so doubtful that it is best to ignore it." (*Nature*, June 27, 1931, 965).

10　"a major event in"—*Papers,* 191.

10　"every available bone"—Oakley, *Antiquity,* 380.

11　"many examples of fully"—Weiner, *Forgery,* 23.

CHAPTER TWO:
The Oldest Humanoid

12　The Natural History Museum: the main building in Cromwell Road was completed in 1881, the year before Woodward joined the staff as a young assistant. The room he used as an office when keeper of geology was shown me by the museum's paleontology librarian, Ann Lum, to whom I am grateful.

13　"a sort of Jules Verne"—Dawson to Woodward, February 14, 1912, quoted from the original at the museum. For the letter in full see *Papers,* 17. Since the two men were together only days before, the obvious question arises as to why Dawson at that time didn't mention the gravel bed and the piece of skull. There must have been a good explanation, however, perhaps given to Woodward in a conversation left unrecorded.

13　Dawson's visit to Conan Doyle: Dawson to Woodward, November 30, 1911, *Papers,* 14.

13　"I think portion"—quoted from the original, see also *Papers,* 17. In the letter the word "think" is unclear, and some prefer to render it as "thick," making the sentence read, "I have found a thick portion," etc. But the word is definitely "think," for it occurs in this same letter in exactly the same form just six lines above, where from its context it is unmistakable. How Dawson could have bungled this particular sentence—in a way the most crucial he ever wrote—becomes an intriguing question. Perhaps what he meant to write was, *I think I have found a thick portion,* and was distracted by the similarity of *thick-think.* Psychological pressures, prompted by the large lie he was telling, would have added to the momentary slip.

14　"leave all to you"—Dawson to Woodward, March 28, 1912, quoted from the original. For the letter in full see *Papers,* 14.

14　"How's that for Heidelberg!"—Dawson, *Naturalist* (1913), 76, and Dawson to Woodward, May 23, 1912, for the timing of the visit, quoted also in *Papers,* 22. The precise number of fragments Dawson had with him on this visit cannot be said. My calculations make three most probable, though it may have been four, certainly no more than that. Neither Dawson nor Woodward left any account of what was said about the skull at this meeting, though they would as a matter of course have talked over the circumstances of the discovery. My description of this is taken from Dawson's subsequent reports.

15　Dawson's background: there is no formal biography, and the bulk of the information about him comes from his obituaries. 1) *Sussex Daily News,* August 11, 14, 18, 1916; 2) *British Medical Journal,* August 19, 1916, 265;

3) *Geological Magazine*, 3 (1916), 477–9; 4) *Dictionary of National Biography* (1951–60), 43. His marriage was reported in *Sussex Daily News*, January 7, 1905.

16 "one of those restless"—Woodward, *Earliest*, 5.

16 Dawson and Barkham Manor: from his deceased law partner in Uckfield, F. A. Langhams, Dawson inherited the position of "steward" not only for Barkham Manor but for Maryon-Wilson's two other holdings in the area, Netherall Farm and the Tarring Camois estate. Dawson and Maryon-Wilson were well acquainted, the latter for long serving as a district magistrate, and Dawson for years being clerk to the Magistrate's Court.

17 For Dawson's telling Maryon-Wilson a "small lie" about his operations at the pit, and the resulting difficulties, see *Papers*, 48–51. Of course, if Dawson misled the manor's owner, then as a matter of course he must have done the same with the tenant, Robert Kenward, whose permission he also required. This fact has pertinence in judging the recollections of Kenward's daughter, Mabel: see below, 256.

17 Venus Hargreaves: about the excavation's main shovel man nothing is known aside from his being a local, recruited by Dawson. He is included in several photographs of the pit, which show him to have been past middle age. He died in 1917 (*Papers*, 24).

17 Teilhard de Chardin: for his first acquaintance with Dawson see his *Letters from Hastings*, 47–8, 53, 58. Dawson's first mention of Teilhard is in a letter of January 15, 1910, *Papers*, 7–8. For the priest's background I have used the biographies by Speaight and Lukas, and other sources as noted.

18 "a very thick, well-preserved"—Teilhard, *Hastings*, 190. This visit and the showing of the skull fragments is also mentioned by Teilhard in a letter of May 18, 1912, to a friend (quoted in Schmitz-Moorman, "Conspiracy," 9), where he adds, "The skull is certainly very curious, of deep chocolate colour and especially of a stupefying thickness . . . unfortunately the characteristic parts, orbits, jaws, etc. are missing." Also for both letters, *Papers*, 20.

18 "with the enthusiasm"—Teilhard, *Hastings*, 198, also quoted in *Papers*, 23.

18 For Teilhard's return to France in July 1912 see his *Letters from Hastings*, 205–6, and Speaight, 45, 47.

19 "It was not until"—Dawson, *Naturalist*, 77. That the jaw was found on Saturday, June 23, is my own calculation from the evidence, but possibly it may have occurred the following day. The fact that Teilhard was not present on that occasion assumes some significance.

19 "was exploring some"—Woodward, *Earliest*, 11. It is well to remember that though Woodward's book was not published until 1948, its first chapter ("The Story of the Discovery") was written no later than 1915.

20 For Woodward's finding of the occiput fragment see his book, *Earliest*, 10. Dawson's description of this find, in one source, appears to say that it

was found *in situ*, rather than in a spoil heap: "Dr. Woodward also dug up a small portion of the occipital bone of the skull from within a yard of the point where the jaw was discovered, and at precisely the same level." (*QJGS*, V. 69, 121). But this is only a clumsy—and misleading—way of saying that the spoil-heap material that yielded the fragment came originally from the pit's lower level. In *Naturalist*, 77, Dawson writes nothing about the level at which the piece was found.

20 "a heap of soft material"—Woodward, *Earliest*, 10.

20 One flint found by each: see the tables of finds in *Papers*, 41, 98, 118. It is not possible to assign specific days for all the animal and flint discoveries, no excavation diary having been kept. Most, it is known, turned up during 1912.

20 "your wonderful find"—*Papers*, 26.

21 "I have been thinking"—Ibid., 27.

21 "Conan Doyle has written"—Ibid., 34. Dawson's final reference to the author occurs in a note of July 1913 to Woodward: "If Banbury can't come I will try and get hold of Conan Doyle for Saty." (*Papers*, 71). The hope was to have the use of Doyle's automobile. Doyle's own letters to Dawson have not survived.

21 Among the early breaking newspaper stories was one in the London *Times*, November 23, 1912, which said in part: "Excavations in Sussex undertaken by an anthropological student have brought to light fragments of a human skull, which are now being pieced together. The skull is said by experts to be that of a palaeolithic man and the earliest undoubted evidence of man in this country."

22 "a glimpse of that"—*Papers*, 28. See also Keith's *Autobiography*, 323–5. Same for the other Keith quotations in this paragraph.

23 "It was late in the"—Keith, *Autobiography*, 325.

23 Keith's diary, December 2, 1912: in a rather difficult scribble the entry begins: "Went down—on way to Zool. tea to Nat Hist Museum to see Smith Woodward's remains—early Pleistocene: of which so many rumours have reached me of late days: Got dark when I worked in hall: lights being turned out. Taken first into his room where he unlocked from his drawer (cabinet) small box with fragments. Biggest fragment an almost complete left parietal . . ." (quoted from the original at the Royal College of Surgeons, London). The remainder of the entry lists the separate specimens and compares lengths, thicknesses, etc. Usually overlooked is the fact that two weeks later Keith paid a second visit to the museum to study the Piltdown remains. In his diary under date of December 15, 1912, he has this: "Been at South Kensington again seeing Sussex skull. Puzzled about jaw." (from the original).

24 "never been exceeded"—*Saturday Review*, December 21, 1912.

24 "several years ago"—*QJGS*, V. 69 (1913), 117. Same for the quotations in the next three paragraphs.

25 "Many years ago, I"—Dawson, *Naturalist* (1913), 75–6.

27 "four years ago"—London *Times*, December 19, 1912. Also London *Daily Chronicle*, December 19, 1912.

27 "a somewhat deeper"—*QJGS*, V. 69 (1913), 121. Same for the next three quotations.

28 "animal remains derived"—Ibid., 123.

28 Woodward's part in the evening is reported at length in *QJGS*, V. 69 (1913), 124–44 (includes illustrations).

30 Smith's part in the evening: Ibid., 145–7.

31 "was human in practically"—Ibid., 150. Same for the next quotation.

32 "a chimpanzee foot"—Waterston, "Mandible," 319.

33 "Our discovery confirms"—London *Times*, December 20, 1912.

CHAPTER THREE:
Challenging the Skull

34 "a grave blunder"—Keith in a letter of July 29, 1913, *Papers*, 75. He says he spotted the blunder "two months ago," on first viewing the skull cast.

35 "I saw Keith and he"—Smith to Woodward, July 4, 1913, *Papers*, 71. Smith further urges that a statement about the correction should be issued "as soon as possible . . . What I fear may happen is that Schwalbe or some other continental authority may arrive at these conclusions before any British statement has been made"—a reminder of how deeply national pride was involved in the Piltdown story.

36 For the July meeting between Woodward and Keith, see Keith, *Autobiography*, 326, and Spencer, *Forgery*, 65.

36 "If Dr. Smith Woodward"—*Times*, August 11, 1913.

37 "*Homo piltdownensis*"—Keith, letter in *Times*, August 14, 1913. Also, Spencer, *Forgery*, 67.

37 "This groove has been"—*Times*, August 12, 1913. Same for the next four quotations. On his assertion that skull reconstruction should proceed as if it were "a problem in Euclid," Keith was quickly challenged. With his ready assent, two fellow scientists broke up a skull, discarding much of it and submitting the remainder to Keith for reassembling. This he accomplished within a few days, and "in the main it answered to the original" (*Autobiography*, 327). He described the feat in *Journal of the Royal Anthropological Institute*, V. 44 (1915).

39 Excursion of geologists to Piltdown: *Sussex Daily News*, July 14, 1913. This lengthy and detailed account gives the names of many who attended, among them Arthur Keith. The Barkham Manor site is described as "the famous little gravel pit alongside the farm roadway, a shallow little pit, scarcely four feet deep . . . the veritable Mecca of all good geologists and palaeontologists throughout the world."

40 Return of Teilhard to England: Teilhard, *Paris*, 93, 95, 98–9; Speaight, 52–3.

40 "While our labourer was"—*QJGS*, V. 70 (1914), 85. See also Woodward, *Earliest*, 12. For Mrs. Woodward's "slender fingers" see *Papers*, 233.

41 "never left us alone"—Teilhard, *Paris*, 104. The letter was written from Jersey on September 10, 1913.

42 "All the debris within"—Dawson, *Naturalist* (June 1915), 182. In this paper about the canine he says that it was found by Teilhard, who "assisted us for a few days," and he then goes on to discuss the tooth itself. It is much larger than in modern man, he says, and "is almost identical in form with that shown in the restored cast."

42 "mapped out in squares"—*QJGS*, V. 70 (1914), 85.

42 "this time we were"—Teilhard, *Paris*, 104–5.

42 "For some time we had"—Woodward, *Earliest*, 11.

43 "We searched for the"—*Evening News* (London), March 3, 1924.

43 "He had me write my"—Teilhard, *Paris*, 105. The tablecloth, containing many signatures afterward embroidered into the fabric, is preserved at the American Museum of Natural History in New York.

44 "one of my brightest"—Teilhard to Oakley, November 28, 1953, *Papers*, 212.

44 "of tremendous importance"—*Daily Express*, September 2, 1913.

44 "corresponds exactly"—*Geological Magazine* (October 1913), 433–4. Same for the quotations in the next two paragraphs.

45 "such as is seen in"—Keith, *Bedrock* (October 1913), 447. Same for the quotations in this and the next three paragraphs.

46 The Lyne report: Lyne, "Radiographs," (1916).

46 For the running exchange during August–November 1913 between Keith and Smith see *Nature*, August 21; September 25; October 2, 16, 30; November 6, 13, 20. Smith did not soften with the years, as late as 1931 giving a Keith book a very negative review (*New Discoveries Relating to the Antiquity of Man*, in *Nature*, June 27, 1931). Among other strictures, he said the book was of no interest "to the serious student," adding blithely that Keith had "never shown much respect for the commonly accepted principles of biology."

46 "I had to speak straight"—*Papers*, 106. Same for the next quotation.

47 "I did not mince"—Keith, *Autobiography*, 327. Same for the next quotation.

47 "singular"—*QJGS*, V. 71, 144.

47 "I was watching"—Woodward, *Earliest*, 44. Also, for the finding of the bone implement, *QJGS*, V. 71, 144–9, and Dawson, *Naturalist* (June 1915), 184.

48 "thrown there by"—*QJGS*, V. 71, 144. Same for the next two quotations (148). The notion of a "cricket bat" was first mentioned in the discussion by Reginald Smith: "he could not imagine any use for an implement which looked like part of a cricket-bat." (148).

50 "It has been suspected"—Gregory, "Dawn," 190.

50   "in Eoanthropus we"—Lankester, *Diversions*, 284.

50   "the earliest specimen"—Keith, *Antiquity*, 353.

52   "obvious incompatibility"—Miller, "Jaw," 3.

52   "the latest ROT"—*Papers*, 134.

52   "a feeling of irritation"—Pycraft to Miller, May 7, 1917, *Papers*, 147. Pycraft excuses himself by saying, "You were so dogmatic, and there was that in your attack which breathed not a little of contempt, which none of us took kindly." The remark nicely illustrates the fact that a certain portion of the Piltdown debate was driven by personalities, which in turn helped deflect any possible thought of fraud.

52   "woefully misread"—Pycraft, *Progress*, as quoted in Spencer, *Forgery*, 103. Same for the next two quotations. Miller's second paper appeared in 1918 in the *American Journal of Physical Anthropology*, running for almost forty pages with bibliography and illustrations. Especially with this lengthy Miller-Pycraft exchange—involving many intensive hours of laboratory study and bringing to bear in great detail all the anatomical expertise of the two dedicated scientists—can the awful and utter waste of energy associated with the fraud be appreciated. But the true total of such wasted effort in the Piltdown story is many dozens of times this particular exchange.

CHAPTER FOUR:
The Sheffield Park Find

54   "But it was evident"—*Sussex Daily News*, August 11, 1916. Same for the next quotation.

54   That Woodward and Smith visited the ailing Dawson in July 1916 is my own conclusion from the fact that the two spent a full two weeks digging at Barkham Manor that month (see *Papers*, 140). Surely at some point they would have taken time to make the short trip to Lewes to see their colleague, then confined to bed.

55   Dawson's death certificate: *Papers*, 136. Inevitably there has been speculation that Dawson's death may have been the result of something other than complications over what has been taken to be aplastic anemia. "While his untimely death remains something of a mystery," writes Spencer, "there does not appear to be any evidence to support the notion that he died of unnatural causes" (*Papers*, 136). But Dawson's death can be called untimely only in the sense that he was comparatively young, and because it complicates matters for those now investigating the fraud. Aside from the newspaper reports and the death certificate, the only evidence bearing either way on the matter is a letter of Helene Dawson in which she confirms the lingering nature of the fatal sickness: "My dear husband's death came at the end of seven weeks' very trying illness, and I was so worn out physically that it was quite easy to get nerve shock" (letter of October 12, 1916: see next entry).

55   "the best and kindest"—Mrs. Dawson to Ruskin Butterfield, October 12, 1916, quoted from the original at the Hastings Museum. This is one of ten letters from Mrs. Dawson to Butterfield, then curator at the museum. All are concerned with the sale of artifacts left by Dawson, and say nothing of Piltdown or other fossils. Also up for sale, she announced, was the Dawson house, Castle Lodge, price four thousand pounds. (In 1993 the house was again on the market, price 350,000 pounds. It has been in the hands of the present owner since 1926.)

55   "that quiet, unassuming"—*Sussex Daily News*, August 18, 1916. Same for the next two quotations.

55   "a genial presence"—Woodward, "Dawson," 477.

55   "a splendid type of"—Keith, "Discovery," 265.

56   "as a boy of twelve"—*Sussex Daily News*, August 11, 1916.

56   Dawson's funeral: *Sussex Daily News*, August 14, 1916.

57   "I believe we are in"—*Papers*, 119.

57   "I am sorry I shall"—*Papers*, 127. A full-size photo of this important postcard appears as the frontispiece to *Papers*.

58   "the recent discovery"—Lankester, *Diversions*, 284. Only "fragments" of the Piltdown cranium were shown to Lankester in 1915 by Dawson, one of them "a part of the frontal with supra-orbital ridge" (Lankester in a letter of December 1915, *Papers*, 135). He did not see the molar. Dawson's purpose in showing these pieces to Lankester remains obscure (there was no direct link between the two, no personal association). It may have been done in an effort to pressure the dilatory Woodward into announcing Piltdown Two. See above, 56–59.

58   "told me that he found"—*Papers*, 163.

58   Netherall Farm: this locale was first mentioned by Woodward in a letter of October 1924, quoted in *The Geology of the Country Near Lewes*, by Osborne White (1926), 67. The fact that, like Barkham Manor, it was also the property of Maryon-Wilson, and was also under the stewardship of Dawson, may be what attracted Woodward's attention to it. Netherall Farm, which still exists, is now an integral part of the Piltdown story, but since the Piltdown Two remains were never actually in the ground anywhere, nothing more need be said of it.

59   "Dawson would not"—*Papers*, 239.

59   "One large field"—*QJGS*, V. 73 (1917), 3. Same for the next three quotations.

60   "few precious fragments"—Osborn, "Dawn," 581. Same for the next three quotations. The photograph of Woodward and Osborn at Barkham Manor appears on p. 589 of this article, also in Spencer, *Forgery*, 108.

61   Barcombe Mills first mentioned: *Papers*, 28.

62   "I have picked up the"—*Papers*, 70.

62   Barcombe Mills entry in museum registry: *Papers*, 187–8, Spencer, *Forgery*, 186, 237.

63 "Smith Woodward probably"—*Papers*, 188. Manuscript report after a visit to the museum by Broom in the spring of 1949.

CHAPTER FIVE:
Curtain Rising

64 The London conference of 1953 and Weiner's reasoning on Piltdown: Weiner, *Forgery*, 26–35. Also see a later statement by Weiner in *Antiquity*, March 1983, 46–8, offering further detail on the steps that led to his suspicions. He clearly states that it all began for him with Oakley's Oxford talk in November 1949 on the fluorine tests: "struck by the anomalous nature of the results he announced . . . I raised it in public discussion (as I recollect) and then went up and spoke to him privately." In the interim before 1953 Weiner considered using X-ray crystallography to test the bones, which he discussed with various experts, but decided he was not well enough trained for the task.

65 "The fact is, all we"—Weiner, *Forgery*, 26. Same for the next three quotations, 26, 29–30, 32. For the postcard and the "earlier letter" see above, 57, and *Papers*, 119, 127–8. Actually, the postcard was not available to anyone for the entire decade of the twenties. On retiring from the museum in 1924 Woodward took many papers with him, including this postcard, apparently by mistake. When he returned it to the museum in 1933 he stated, "I think that this should be carefully preserved" in order to quiet "doubters for whom Dawson's own record is needed" (*Papers*, 128).

67 "apparently no more"—Oakley & Hoskins, "Evidence," 381. Here again is one of those minor puzzles that dot the Piltdown story, this one centering on the tendency of the experts to overlook things. Oakley adds that the fact of the white dentine beneath the very thin surface stain was surprising—was revealed "most unexpectedly"—and he pointedly compares the condition with "recent teeth." He even observes that the fluorine test results "scarcely provide any differentiation between *Eoanthropus* and recent bones," at the same time conceding that this "requires some explanation" (381). Yet with that, writing in February 1950, he simply drops the subject. Perhaps some of this was at the back of his mind when he received Weiner's telephone call in August 1953 (see above, 68) and so rapidly agreed with the idea of forgery. Of course Weiner also at first missed the implications of the white dentine, since it may be accepted that he read Oakley's *Nature* report.

68 "You don't really mean"—*Antiquity*, V. 57, 47. Same for the quotations in the next two paragraphs.

69 Details of the fraud exposure are given in two bulletins of the British Museum (Natural History): *The Solution of the Piltdown Problem*, by Weiner et al., 1953 (six oversize pages, illustrated) and *Further Contribu-*

tions to the Solution of the Piltdown Problem, by Weiner et al., 1955 (sixty-five pages, illustrated). Also Weiner, *Forgery,* 36–78. For some reason, Weiner does not make clear in his book that the investigation proceeded in two phases, as my chapter shows.

69 "unlike that found"—Weiner et al., *Solution* (1953), 142.

69 "the occlusal surface"—Ibid.

70 "tough, flexible"—Ibid., 144.

70 "a most elaborate"—Ibid., 145.

70 "so extraordinarily"—Ibid.

71 "it is opportune to"—*Times,* November 24, 1953.

71 Visit of Weiner and Oakley to Keith: both men made written reports, which were filed at the Natural History Museum; see *Papers,* 207, 219–21. The remarks quoted for the visit are from these two reports.

72 "The fragments of the"—*Times,* November 21, 1953.

74 "incontestably the"—Weiner et al., *Contributions,* 259. The idea of thickening bone artificially by chemical action was also offered as a possible solution. In one test a skull fragment was soaked for a week in a mixture of potassium hydroxide and tap water. The resulting increase measured more than a third of the original thickness, with most of the expansion in the *diploe,* which matched Piltdown. The experimenter, Ashley Montagu, stated that with more patience and skill, "one could do a great deal better than that." Kenneth Oakley demurred, however, saying that subfossil bones, lacking sufficient collagen, could not be appreciably expanded, and the uniform structure of the Piltdown *diploe,* the regularity of its tiny spaces, denied the possibility of an artificial increase. For the test, and commentary, see *Nature,* July 9, 1960, 174.

74 smell of burning: Ibid., 255–6.

76 "Present thinking"—Washburn, "Hoax," 761–2. Interestingly, Washburn, who was at the 1953 London conference, came close to having a direct hand in the exposure. At the banquet (see above, 64) he sat with Weiner and Oakley and had the same surprised reaction to Oakley's revelations about not knowing the site of Piltdown Two. His strong suspicions aroused about fraud, on his return to Chicago he made some experiments, Weiner and Oakley cooperating with him by mail. But apparently at some point it was decided, with Washburn's concurrence, "to keep the exposure an entirely British affair" (*Papers,* 201). For the relevant information on Washburn's brief involvement see Weiner, *Forgery,* 26, 49; *Papers,* 199, 201, 214; Spencer, *Forgery,* 133, 229; and Washburn in *Current Anthropology,* June 1992, 276 (the Tobias article).

76 "I do not agree"—Hooton, "Affair," 288.

77 "congratulate you"—*Papers,* 212. Same for the quotations in the next paragraph.

78 "Sore subject!"—*Papers,* 248, quoting Oakley's manuscript notes of the meeting. Same for the next two quotations.

79 "I am frankly puzzled"—*Papers*, 249.

79 Carbon-14 dating of the Piltdown specimens: *Nature*, July, 25 1959, 224–6. The tests were conducted by Oakley and H. de Vries. Same for discussion of the orang skull's source. For some later refinement of the age question see the report in Hedges, et al.

81 Weiner's 1953 visit to Lewes and vicinity: Weiner, *Forgery*, 154–88, as enlarged on in his letters in *Papers*, 216–18. He does not provide a consecutive record of his movements but they are readily reconstructed from what he does say.

83 "The Council, in the"—*SAC*, 1904 (*Report* for 1903), xiv–xv. See also *SAC*, 1905 (*Report* for 1904), xv–xvi; *SAC*, 1946, 85; and Weiner, *Forgery*, 174–5. Doubts have been voiced about Dawson's action in the purchase of Castle Lodge being the underhanded maneuver pictured. But they all stem from misunderstanding of a letter of Louis Salzman to the *Times* (January 23, 1955) correcting Weiner's statement that Dawson, in buying the lodge, had been asked by the society to act in its behalf (stated in *Sunday Times*, January 16, 1955). Salzman says that Dawson was not so requested, and this has been taken as a defense in general of Dawson in the matter. But Dawson did make unauthorized use of the society's notepaper, and he allowed the seller to assume an official link with the group. In 1946 Salzman stated clearly that Dawson "was a prominent member of the Society, though not on the council, and the vendors seem to have believed that he was buying the house on behalf of the society. The blow was entirely unexpected" (*SAC*, V. 85, 38). A correspondent of Salzman in 1955 recalled "the anger of the members" at the news that the property had been sold "in such a manner" (Stretton to Salzman, November 4, 1955, Salzman Papers, SAS).

84 "the daily coolness"—Weiner, *Forgery*, 175.

84 "but little judgment"—*SAC*, 1946, 282. Same for the next quotation.

85 The Morris cabinet and notes: Weiner, *Forgery*, 154–61, and *Papers*, 218. A photograph of the annotated flint is in Spencer, *Forgery*, 142. The physical description of the Morris notes in Weiner and *Papers* is somewhat misleading as to their number and disposition. My own description rests on study of the original notes preserved in the library of the Natural History Museum in London (Piltdown papers, DF 116/117). The quotation from *Macbeth* bears the notation "Act I 3," but the correct citation is I, 4:51–53. The first note is written on a standard 3×5 index card. The second is on what appears to be a piece of photographic backing, 4×5 in., with the writing done on a slant. The words "Watch C. Dawson. Kind regards" were certainly written at a different time from the longer message inked over them, and the two do not appear to be linked.

87   "We had one episode"—*Papers*, 216. The London *Daily Mail* (January 6, 1955) heard about the skulls supposedly found at Castle Lodge and sent a man to investigate. He talked with Nicholl at length but found little more than had Weiner. This time, however, it was said that the skulls were found not under the floor but in a cavity in a wall behind "wainscotting near the window bay." No more was ever heard of these skulls, nor was Castle Lodge ever searched by Weiner.

87   "Who did it? is a"—*Times*, November 21, 1953. Same for the quotations in the next paragraph.

88   "Charles Dawson was"—*Times*, November 25, 1953.

88   "was not the type of"—*Daily Herald* (London), November 23, 1953. The account adds that the brother, Trevor, had come to London from Scotland "to consult members of the family" in the matter.

89   "he was nearly stone"—*Papers*, 231, also 232–5, 241. Unpromising as Essex was as a witness, he has since managed to gain some further attention. See especially Bowden, Blinderman, and Spencer.

89   "several pieces of bone"—*Papers*, 240. See also Weiner, *Forgery*, 95, 188.

89   The St. Barbe-Marriott testimony: Weiner, *Forgery*, 164–7, and *Papers*, 225, 229–31. The several quotations are from both sources. For Dawson's quite open use of chemicals at the time of these incidents, see *Papers*, 76, where he asks Woodward to supply some "hardening solution . . . made of spirit and shellac . . . I am quite out of it." At this time he was hardening a fossil elephant tooth. In a letter the previous month he asks Woodward for a gelatinizing recipe for bone preservation, and adds, "I have got a saucepan and gas stove at Uckfield," meaning in his office (*Papers*, 71).

90   "Their hesitancy is"—Weiner, *Forgery*, 166.

92   "It is not merely"—Ibid., 185.

93   Weiner on Dawson's background: Ibid., 171–88. For his theory about the planting of the skull, Ibid., 192–3.

94   "clearly dissociate"—Ibid., 199.

94   "unknown manipulator"—Ibid., 200

94   "We have seen how"—Ibid., 203. Same for the next quotation in the paragraph.

94   "In the circumstances"—Ibid., 204.

94   Weiner's lectures on Piltdown: only one of these has been published, partially, as an appendix in King, *Teilhard*, 159–68. Delivered April 28, 1981, at Georgetown University, Washington, D.C., it defends Teilhard against Gould's charges. As late as 1979 Weiner stated his firm belief that Dawson was the sole culprit at Piltdown.

CHAPTER SEVEN:
List of Suspects

95   For some extreme examples of tangled and weblike conspiracy theories, see the articles by Matthews, *New Scientist*, 1981, and Thomson, *Ameri-*

can Scientist, 1991. Apparently not everyone is bothered by such excesses. The rather startling convolutions offered by Matthews in *New Scientist*, for example, make an appearance, unchanged and unquestioned, in Broad & Wade, 119–21, where they are presented as if fully proved.

96   The Butterfield case: first presented in Von Esbroeck, *Lumière*, 1972. For recent discussion see Blinderman, 117–20, and Spencer, *Forgery*, 165–7.

96   The Abbott case: first fully presented in Blinderman (1986), 191–218. For a firsthand look at the quirky Abbott, and initial suspicions about him, see *Papers*, 36–7, 74–5, 100, 101, 222–3. For direct contact between Abbott and Dawson, see *Papers*, 24, 33; also Weiner, *Forgery*, 96–150, *passim*.

98   The Sollas case: first reported by L. B. Halstead in *Nature* in 1978 (V. 276, 11–13; for some follow-up, see V. 277, 596). For more recent discussion of the Sollas thesis see Blinderman, 183–5, and Spencer, *Forgery*, 168–72. For an example of the newspaper coverage of the Sollas charge see *Sunday Telegraph*, October 29, 1978.

99   "I know that Sollas"—Halstead, "Light," 11.

100 The Smith case: first reported in Millar, 226–37, but it has attracted little support. For some discussion see Blinderman, 219–31, and Spencer, *Forgery*, 172. Ian Langham's support for the Smith case (*Artefact*, 1978) was later withdrawn when Langham switched his suspicions to Sir Arthur Keith. See above, 151.

100 Smith's Australian fossil, the Talgai skull, had a short life indeed, but it was referred to briefly by both Dawson (*Naturalist*, 1915, 183) and Woodward ("Early Man," 1917, 4).

100 "I was first inclined"—Millar, 234.

101 The Hinton case: first openly suggested by Halstead in the *Times*, November 25, 1978, and in *Nature*, February 1979, 596, and expanded in Bowden (1981) and Matthews, *New Scientist* (1981). For recent discussion see Blinderman, 145–53, Spencer, *Forgery*, 175–8, and Thomson, "English" (1991). Hinton's letter to the *Times* appeared on December 24, 1953. For some corroboration of his claim in this letter about early suspicions, see his letter of 1916 (*Papers*, 137), where he talks of *Eoanthropus* eventually turning out to be "only a bad dream."

101 "Had the investigators"—*Times*, December 4, 1953.

102 The Woodhead-Hewitt case: first presented in 1985–6 in Costello, *Antiquity*, LIX, 167–73 and LX, 145–7.

103 "found more pieces"—*Antiquity* (1985), 169.

104 "A small fragment of"—*QJGS*, V. 69 (1913), 121.

104 "ventilated many"—*Antiquity* (1985), 167.

105 *Probability* as a technique: the real drawback, the essential flaw, in the overuse, and even at times the mere use, of probability in argument (better identified as *supposition*) is that it cannot be logically refuted, only

summarily denied. Ultimately this leaves any discussion or disagreement precisely nowhere. Its inherent weakness and loss of force in cumulation is another matter.

105 "Is it possible that"—Grigson, "Links," 58. Not surprisingly, the charge against Barlow rests entirely on the old idea that Dawson, as Grigson phrases it, "did not have the expertise to carry out the whole range of brilliant forgeries" at Piltdown, but would certainly "have needed help." She does claim to have "a new piece of evidence," relating all too murkily to some visits Dawson paid to the Royal College of Surgeons in the fall of 1913 (some two months after the canine was found). How these few doubtful facts were seen as adding up to a charge of complicity against the unoffending Barlow prompts only wonder. Grigson's liking for summary judgments can be seen in what she says of Sir Arthur Keith, whom she never knew personally. She admired him, she says, for some aspects of his work, but also thought him "a conceited, humourless bore, with very little knowledge of archaeology" (*Current Anthropology*, June 1992, 266). This contrasts with the figure recalled by his friends as "a much-loved man, kindly and gentle in manner, friendly and unassuming, of a somewhat retiring disposition" (*Dic. Nat. Bio.*, 1951, 565–66).

CHAPTER EIGHT:
The Writer Accused

107 Doyle's last hours and death: Nordon, 158, 169; Carr, 281–2; Higham, 334.

108 Doyle and spiritualism: Hall, 91–143. See also the biographies by Carr, Higham, and Nordon. Hall correctly notes the general reluctance, especially by his biographers, "to acknowledge the fact that the last fourteen years of Doyle's life were dominated by his devotion to spiritualism" (91).

108 "will mark an epoch"—Doyle, *Fairies*, 39. He adds that in time "These little folk who appear to be our neighbors, with only some small difference of vibration to separate us, will become familiar. The thought of them, even when unseen, will add a charm to every brook and valley . . ." (57). His real hope, as he frankly stated, was that acceptance of the reality of the fairy world would open the way for acceptance of spiritualism.

109 "turns up a surprising"—Winslow, 33.

109 "there was another"—Ibid., 34.

109 "malicious"—*Baker Street Misc.* (1987), 29.

109 "a cheap publicity hype"—Ibid., 28.

109 "an evidence-free argument"—*Science 83* (November), as quoted in Elliott, "Curious," 11.

110 "obfuscatory of science"—Langham, "Sherlock," 4. Langham ends his article with a veiled reference to "the real mastermind" at Piltdown,

which he says he will name "on a later occasion." He meant Sir Arthur Keith, and the revelation came in 1990 in the Spencer book (see Chapter 10).

110 "Evidence has to be"—Elliott, "Curious," 32.

110 "is simply not true"—*Baker Street Misc.* (1987), 35. Same for the next quotation.

111 Among the items of possible "evidence" missed by Winslow which support the idea of Doyle's "guilt" is the whole question of Holmes' abiding interest in *chemistry* (and by extention Doyle's), so crucial a factor in the Piltdown fakery. In the tales the detective is shown to be a "first-class" chemist, making many abstruse experiments. For a comprehensive look at Holmes-as-chemist see "The Chemical Corner" by C. O. Ellison, in Hall, *Creator*, 30–38.

111 Doyle's 1909 museum inquiry: *Papers*, 5. The inquiry, sent to Woodward, was relayed to Dawson, who replied to Woodward, "I am interested in what you say about Crowborough and the footprints . . . I shall be very pleased to make Sir A. Conan Doyle's acquaintance. If I am in Crowborough first I will call on him or if he comes to Lewes I hope he will come and see me." Nothing indicates how the meeting finally occurred, except that Dawson may be the "expert" referred to by Doyle in Nordon, 329: see the next note.

111 "I have another expert"—Nordon, 329. The expert is not identified, but it may have been Dawson, acting at the request of Woodward: see the previous note.

111 Woodward's 1911 visit to Windlesham: *Papers*, 12, in Dawson to Woodward, May 13, 1911. Nothing is said of the date or purpose of the visit.

111 "My wife and I went"—*Papers*, 14.

112 "A sort of Jules Verne"—Ibid., 17.

112 Doyle's visits to Piltdown: *Baker Street Misc.* (1987), 37, and Weiner, *Forgery*, 105.

112 "Doyle was also a"—Winslow, 36. Same for the next three quotations.

115 Holmes tales: for the quotations from *Hound, Garridebs,* and *Devil's Foot* see Baring-Gould, II, 7, 22, 52, 54, 514, 515, 647. For *Pithecanthropus* in *Lost World* see 170, Chap. 13, and 219, Chap. 16 (Crowborough Edition).

117 "A tall, thin, bitter"—Doyle, *Lost World*, 54, Chap. 5. Page references, here and below, are to the Crowborough edition of the *Complete Works*.

118 "of an oval contour"—Ibid., 144, Chap. 6.

118 "intriguing resemblance"—Winslow, 40.

118 "vamped up for the"—Doyle, *Lost World*, 45, Chap. 5

119 "Lankester had arranged"—Winslow, 41. Same for the next two quotations.

120 "Doyle was a sportsman"—Ibid., 42–3.

120 "Such gullibility must"—Ibid., 43.

121 "an enormous variety of"—Weiner, *Forgery*, 108. Weiner went to visit the Gerrard firm and was told by the owner himself that "unmatched

jaws and other odd bones were probably easier to come by in the years before World War I than now . . . ape and human jaws could be easily come by and many geologists had them."

121 "I've seen them in the"—Doyle, *Lost World*, 121, Chap. 10.

121 "an area, as large"—Ibid., 40, Chap. 4.

122 "The Piltdown hoax was"—Winslow, 39. The idea that Doyle, if he was the hoaxer, might have been seeking publicity for his forthcoming novel is not worth lingering over. He would have been disappointed in any case, for the press made not the least connection between Piltdown and the book.

122 "I leaned my head"—Doyle, *Lost World*, 142, Chap. 11.

123 "Since most of the"—Winslow, 36.

124 Discovery days at Piltdown: there is no running record of the Piltdown excavations (one of Woodward's grievous oversights), formal or informal, so compiling totals has been a matter of combing all sources for dates and occasions. The effort was made infinitely easier by the tables, with their notes, in *Papers*, 41, 98, 118.

126 "For answer the professor"—Doyle, *Lost World*, 34–5, Chap. 4. Lankester had another, even more direct link to the novel. In a letter to Doyle discussing the story (August 1912) he suggests including "a gigantic snake sixty feet long." Doyle used the idea, reducing the length to fifty-one feet (Nordon, 330).

127 The Doyle-Piltdown question has not escaped a measure of the standard Sherlockian foolery, with Holmes himself named as coconspirator. See *Baker Street Journal*, V. 13, No. 3, 150, and a reply in V. 15, No. 1, 28–31. Also *New Statesman*, November 27, 1954. Another famous author who surfaced briefly in Piltdown matters was Rudyard Kipling (died 1936). A resident of Sussex, living not far from Barkham Manor (Burwash, where he occupied a splendid old house on extensive grounds), Kipling was also active in the affairs of the Sussex Archaeological Society. His conjectural link to Piltdown relates to a short story he published in 1928, *Dayspring Mishandled*. Because the plot concerns a forgery (a spurious manuscript), and because the word "dayspring" is poetic for "dawn," both Weiner and Oakley became curious. They thought it possible that Kipling might have had private knowledge of the fraud's secret background, perhaps something in line with his tale's motive of revenge. But the reference led nowhere. (See Weiner, *Forgery*, 118; *Papers*, 249–50; Spencer, *Forgery*, 236.)

CHAPTER NINE:
The Priest Accused

128 Teilhard's departure from England: Teilhard, *Hastings*, 205; *Papers*, 24, 26; Speaight, 44–5.

129 "The very strongest"—quoted in Cole, *Leakey's*, 374. Cole goes on to say that Leakey first hinted at Teilhard's guilt in his *Unveiling Man's Origins*

(1969) and had "drafted part of an entire book on the subject," even though he had "no real evidence, only a hunch." He also did some hinting in his *By the Evidence* (1974).

129 "primal fascination"—Gould, "Conspiracy," 11. Same for the next quotation. Gould's first article on Piltdown (*Natural History*, 1979) showed no serious interest in assigning guilt but probed the question, "Why had anyone believed Piltdown in the first place?" As to the perpetrator, he thought it was "most probable . . . that Dawson acted alone." At the same time he amused himself with the idea of Teilhard and Dawson "over long hours in field and pub, hatching a plot," imaginative speculation that, as he confessed, "provides endless fun." It was only with his second article that he grew serious about Teilhard.

130 "nothing excluded Teilhard"—Gould, "Conspiracy," 14. Same for the next three quotations.

130 "As far as the fragments"—*Papers*, 212, where the letter is given in its entirety. Of the two sentences in the quoted passage, Gould gives only the second.

131 "cannot be"—Gould, "Conspiracy," 16. Same for the other quotations in this paragraph.

131 "Concerning the point"—*Papers*, 236.

132 "planned the"—Gould, "Conspiracy," 200.

132 "a chance meeting"—*Papers*, 212.

132 "In 1908 I did not"—Ibid., 242. Same for the quotation in the next paragraph.

133 "These letters speak"—Gould, "Conspiracy," 21.

133 "Piltdown never again"—Ibid., 22. Same for the next quotation.

134 "glumly walked"—Ibid., 25.

134 "an active coconspirator"—Ibid., 28. Same for the other quotations in this paragraph.

135 "He was a passionate man"—Ibid., 28.

136 "fatal error"—Ibid., 16.

136 "major points"—Gould, "Letters," 27. This is a follow-up article by Gould answering critics of his Teilhard thesis. Nothing of significance is established in it, a fact that was readily admitted by Gould (*Hen's Teeth*, 228).

136 "two foundations"—Ibid., 27.

136 "my single strongest argument"—Ibid., 28.

137 "one large field, about"—*QJGS*, V. 73, 3.

137 Strangely, none of the detailed rebuttals offered to Gould's thesis make even a passing mention of Barcombe Mills. Lukas (*America*, 1981) suggested that in 1913 Teilhard may have been shown a second pit at Barkham Manor, then later she changed that to another pit at Sheffield Park (*Antiquity*, 1983). In this article, Lukas' translations of Teilhard's letters at many points differ materially from those given in Teilhard,

Hastings, 190–1, 197, and Teilhard, *Paris*, 98. Her use of "park," for instance, where the other source has "ground," leads to much confusion. Gould's article is reprinted in his *Hen's Teeth* (1983), with considerable added comment (229–40), none of which does much to advance his argument. First to mention Barcombe Mills as possibly the site visited by Teilhard in 1913 was Spencer, *Forgery*, 186. The references to it in Tobias, "Appraisal," 248, are misleading since the name itself did not enter the discussion until Spencer raised it.

137   Teilhard in England, 1913: Teilhard, *Paris*, 95–103, *Papers*, 76, 79. See also the biographies by Speaight and Lukas.

138   Teilhard to Oakley, 1953: *Papers*, 212.

138   For some sample defenses of Teilhard by his partisans, see *The Teilhard Review*, V. 16, No. 3 (1981) and the two articles by Lukas.

139   "could not have seen"—Gould, "Conspiracy," 16. For a response pointing out Gould's error, see King, *Teilhard*, 161.

139   "eleven letters refer"—Ibid., 20.

140   "It was during the"—Teilhard, *Heart*, 25–6.

141   For the essay "Cosmic Life" see Teilhard, *Writings in Time of War*, where it is noted that the piece was "finished on 24 April 1916 at Nieuport."

141   "deceived and cheated"—Gould, "Letters," 23 (comment by G. H. Von Keonigswald).

142   "of course nobody will"—*Papers*, 212.

143   "The big fragments of"—Schmitz-Moorman, "Hoax," 11, quoting Teilhard to Breuil in December 1953. See also Lukas, 330.

143   "The only thing that"—*Papers*, 212. The sentence immediately preceding the quotation runs: "When we were in the field I never noticed anything suspicious in his behaviour." Since the finding of the two mysterious fragments was not suspicious but only *puzzling*, in Teilhard's mind there was left room for an explanation—but suspicions can also be explained. That there was more behind Teilhard's reaction that went unexpressed is plain from the fact that in the letter he fails to state what constituted the puzzlement. Oakley, who should have asked for speedy elaboration, simply let it go, as has everyone since.

145   "I read this statement"—Gould, "Conspiracy," 25.

145   Some find the thought of Teilhard's continued silence over Piltdown, if he indeed suspected something, to be disturbing morally. It has been suggested that he was somehow bound to silence by his character as a priest, but for that idea there is little or no basis. More probably, he simply could never make up his mind as to exactly what was wrong, and precisely who was responsible and to what extent. While he did not treat Piltdown in his evolutionary writings, he did allow the Piltdown cranium a place in man's descent, rejecting the jaw as that of an ordinary Pleistocene ape (agreeing with Miller). From that very fact it can be legitimately concluded that he *must* have wondered—could not have avoided

it—just how the two unrelated specimens happened to come together so conveniently in the same pit. Even some of Teilhard's followers now concede that he well may have suspected something, yet kept quiet. (See Dodson, "Co-Conspirator," 19, 20, and Lukas, "Haunting," 7, 11.) Weiner, too, in later years felt that Teilhard "had suspected some irregularity" but could never bring himself "to denounce anybody" (King, *Teilhard*, 163).

146   "Spent an enjoyable"—Teilhard, *Paris*, 105. Also quoted in *Papers*, 80.

146   The *Antioch Review* article, by Howard Booher, assumes two factors as its foundation: Dawson's supposed lack of scientific knowledge and the stipulation of legitimacy for the cranium. At this juncture, the first warrants no discussion (but see my Chapter 11). Neither is the second, blithely ignoring the exposure report, entitled to a response (but see above, 72–74). Sufficient comment is Booher's putting implicit faith in the testimony of such biased and unreliable sources as Robert Essex and Mabel Kenward.

147   The Clermont article asks whether Piltdown may have been fabricated as "a practical joke on the 50th anniversary of Moulin-Quinon." Once more, anyone who can view this elaborate fraud as having been done as a joke simply has not absorbed the necessary information. Clermont's argument is further vitiated by several basic errors—for example, that Teilhard was present for the finding of the jaw, which he definitely, on his own word, was not (*Papers*, 212).

CHAPTER TEN:

The Anatomist Accused

149   "In a complex business"—*Times*, January 9, 1955. Same for the next three quotations.

150   "Never was a man so"—Tobias, "Appraisal," 287, quoting from Grigson, "Keith," *Times Literary Supplement*, February 1, 1991.

150   "So compelling was the"—Tobias, "Appraisal," 290, quoting from a letter of Keith to Ashley Montagu, September 1954. In his *Autobiography* (1950) Keith says of Dawson that "his open and honest nature and his wide knowledge endeared him to me" (328). In his *The Antiquity of Man* (1925 ed.) he takes note of Dawson's "sterling ability and unselfish personality" (486).

150   Ian Langham: for Langham's work and his part in the Keith case see the introduction and preface to Spencer, *Forgery*, and Tobias, "Appraisal," 250-1.

151   The Tobias article of 1992 is an impressive witness to the fascination that Piltdown can still exert four decades after the initial exposure. It runs almost twenty oversize pages, is followed by an additional seventeen pages of commentary offered by nineteen leading scientists, and ends with a Tobias rebuttal covering another dozen pages. Its bibliography lists 181 items.

151 For Langham's case against Smith, later abandoned, see his thirty-page article in *Artefact* (1978).

152 "This has been an"—from the original at the Royal College of Surgeons. Also quoted in Spencer, *Forgery*, 188 (where several words have been accidentally omitted, though without altering the sense). I supply opening and closing sentences of the passage, which are not given by Spencer. Incorrectly, Spencer calls it a "weekly" diary (238), but the spasmodic dating of the entries shows otherwise. Among the entries for 1912–14 are a good few references to Piltdown, and to the book manuscript Keith had then in progress, in which he was treating the topic at length (*Antiquity of Man*, 1915). None supply anything of significance for present purposes.

152 "Leaving the question"—*British Medical Journal*, December 21, p. 1,719. Anonymous, but by Keith.

153 "Yesterday Celia and"—from the original at the Royal College of Surgeons, also quoted in Spencer, *Forgery*, 190.

154 "I had never heard"—as quoted in Spencer, *Forgery*, 190, from *The Sphere*, 1913.

155 "that Keith was in some"—Spencer, *Forgery*, 191.

155 "I saw this tall man"—from a taped interview (1973) with Mabel Kenward, as quoted in Spencer, *Forgery*, 238. She had earlier told of the same incident in *Sussex County Magazine* (1955) without the details of the intruder's appearance.

156 "One morning early in"—Keith, *Autobiography*, 328, also quoted in Spencer, *Forgery*, 192, where Langham's reasoning on the point is detailed.

156 "On Thursday went to"—Spencer, *Forgery*, 194, where it is discussed at length. The complete sentence in the diary for that day—squeezed between items for Wednesday and Friday—which is not given by Spencer, reads: "On Thursday went to Brighton Museum Association & Sussex County Hospital to meet Dr. Holehouse about specimens for Museum." Note that he does not mention the Hastings excursion, but does mention having something personal to do besides taking part in the association meeting.

156 "ran an excursion to"—Spencer, *Forgery*, 194.

157 Dawson and the thirteenth vertebra: explained and discussed in Spencer, *Forgery*, 195.

157 "extraordinary enterprise"—Ibid., 199.

158 "a suitable skull"—Ibid., 197. Spencer leaves no doubt of his belief that, as Dawson claimed, a veritable skull was first turned up, in whole or in part, by the men digging for road metal: "This much is certain: a skull, or at least part of one, was uncovered and accidentally shattered by farm labourers while working the gravel bed at Barkham Manor" (197). For strong reasons against both of these ideas see above, 212, and below, 256.

Spencer also feels that the substitution theory of the skulls "is not unreasonable" (198). For a refutation see above, 212–14. Tobias also appears to have missed the rather obvious fact that the specimens from Sheffield Park and Barcombe Mills were never actually in the ground. The many Piltdown specimens, he declares, had to be "prepared, stained, and salted, presumably at different times, into three different localities" (Tobias, "Appraisal," 282, see also 284).

158   "the discussion naturally"—Spencer, *Forgery*, 198. Many or most readers of the Spencer volume, judging by reviewers and critics, do not fully understand that it offers literally nothing to explain how or when Dawson and Keith are supposed to have joined forces. The why is the usual one of the amateur Dawson needing scientific help ("Dawson is most unlikely to have been able to conceive the broad plan and to plot the minutiae of its execution on his own," etc., 259). Emphasizing the point, Tobias quotes (285) another Piltdown commentator, Zuckerman, who insisted that "the faker must have known more about private anatomy than all the highly distinguished anatomists he deluded." But this of course is the conclusion of committed professional scientists reluctant to admit that a clever interloper could gain the necessary specific knowledge on his own, choosing only what he needed, without mastering an entire discipline. The remarkable abilities wielded by a dedicated charlatan are always a surprise to honest men.

159   "master-stroke"—Spencer, *Forgery*, 198.

159   "acting without Keith's"—Ibid., 206. Same for the next quotation in this paragraph.

159   "the man on the spot"—Ibid., 202. Same for the quotation in the next paragraph.

161   Keith at the pit on January 4, 1913: the suggestion that Keith and his wife actually did reach the pit that day, instead of halting at the gate, was first offered by R. V. Wright of Sydney University, in *Current Anthropology*, 1992. Wright was also the first to cite the *Sphere* story of January 18, 1913, with its photo of the flooded pit. (See under Tobias.) In this same connection, the periodic flooding of the pit by rainwater, a passing observation might be registered. Never, apparently, during the years of active excavation, did anyone think to erect over the pit area any sort of protective roofing. A simple wooden structure would have sufficed, allowing work to go on. Yet there is no sign that any such arrangement was ever contemplated, and it is hard to find a reason for the oversight. Failing a broad roof, just the pit itself might have been closed off by a platform, to be laid down and picked up in minutes, but even this was neglected. During the winters, when digging was halted for three or four months, the pit, so far as is known, remained open to the elements, and to the curiosity of every passerby. In 1912, excavation techniques may have been relatively primitive, but not to the point of overlooking com-

monsense requirements. Why in this case one of the most basic requirements *was* overlooked is baffling. Whether this reflects more on Woodward's competence or on the standards prevailing at the time is a nice
question, perhaps a bit of both. In any case, it is of a piece with Woodward's failing to maintain any sort of excavation diary.

162 "was not a member"—Spencer, *Forgery*, 189. If Keith had asked Woodward or Dawson about the location and nature of the pit, a sufficient reply
would have taken a very few seconds, something like "It's a flint quarry
beside the main road on Barkham Manor, just west of Uckfield. About a
mile from the river, on flat land." A similar advance story, rather like
Keith's in its detailed description of the finds themselves, appeared in *Nature* a week before the Geological Society meeting (December 12, 1912,
quoted in Millar, 112–13). Though not mentioning Barkham Manor, it
does give some specifics of location and circumstance, saying that the discovery was made "in a gravel which was deposited by the River Ouse near
Piltdown Common, Fletching, Sussex . . . Although the basin of the
stream is now within the Weald and far removed from the chalk, the
gravel consists of iron-stained flints closely resembling those well-known
in gravel deposits on the downs . . ." The article's information came from
Woodward, who is referred to in its text.

162 "pinpointed several"—Tobias, "Appraisal," 253. Same for the next quotation.

163 "Keith had taken part"—Ibid. Aside from the objective evidence, the in-
trinsic weakness in this paragraph is clearly to be seen in the frequent use
of *probability*, here invoked four times within fifteen lines. Spencer also
offends in the overuse of probability. In the course of only a dozen pages
summing up his case against Keith (199–208, plus a section of notes,
241–2), he employs the device no less than *fifty-eight times* using a wide
variety of phrases: *it is conjectured, every reason to suppose, it is patently clear,
can be little question, may have been, seem to favour, every indication, from all
appearances, tempting to suggest*, etc., etc. See above, 105, for some comment on the use of probability in Piltdown studies.

163 "could be no question"—Spencer, *Forgery*, 194.

163 "The fascinating romance"—*Sussex Argus*, July 15, 1911, quoted from a
clipping found at the Sussex Archaeological Society, Lewes (pasted into
the rear of a volume of Moss' *History of Hastings*). An advance story in the
Argus (July 7, 1911) gives the official reception committee as composed
of nine members, all named, and Dawson is not among them. Neither
does the paper mention that there were any guests of honor.

164 "Before entering the"—*Hastings and St. Leonards Observer*, July 15, 1911
(also from a clipping in Moss' *History of Hastings*, at the Sussex Archaeological Society). Nothing is said here about guests of honor either.

165 "pattern"—Tobias, "Appraisal," 251. Same for the next four quotations,
251–5.

166 "the earliest known"—Keith, *Autobiography*, 645.

167 "in a bonfire some"—*Papers*, 220, quoted from Weiner's notes. Also *Papers*, 207, where Oakley's notes of the visit say much the same. Nothing is stated in either place about when or why the letters were destroyed.

167 "I am very anxious"—*Papers*, 22. Never published, Dawson's original manuscript, a mere eight typed pages, is preserved at the Natural History Museum. Spencer uses the fact that Keith actually reviewed the Le Double book in 1915 (in *Man*, V. 15, No. 37) as a means of linking him with Dawson's paper on the topic (195). I suspect that what really happened was the reverse: when Dawson found that Keith was well aware of the Le Double volume—he praised its "valuable" thesis—he quietly withdrew his own plagiarized paper. Dawson's plagiarism in this instance is no different from his earlier literary thefts (see Chapter 11).

168 A good example of the favorable-but-uncertain response to the Langham-Spencer-Tobias theory is provided by C. B. Stringer of the Natural History Museum. Though accepting the charge against Keith as "the best case yet," he points out two items that tend to support Keith's innocence: 1) the fact that the canine went directly contrary to Keith's cherished ideas, often publicly expressed, on evolution, and 2) the highly detailed, fifty-page study, based on no less than a year's research, that Keith made in 1938 comparing Piltdown with Swanscombe, hardly the action of a forger (Stringer, in Tobias, "Appraisal," 273). A few commentators have thought that the sheer scientific workmanship involved in the fraud was not that of an accomplished professional (contradicting most others). John Langdon, of the University of Indianapolis, has carried this idea to the length of asserting that it proves Dawson to have acted *without* such expert assistance, and therefore alone. The forgery, says Langdon, was "fairly good for its time in terms of palaeontology but weak anatomically. Although some of the discrepancies are very subtle, taken together they appear to belie the hypothesis that an expert in anatomy was involved . . . the forgery appears to be the work of an amateur attempting a serious scientific fraud" (Langdon, "Misinterpreting," 630). No doubt, but a remarkably brilliant amateur!

CHAPTER ELEVEN:
The Wizard of Lewes

169 Dawson's grave: listed in the church's burial record book No. 25 as entry 234. Dawson's funeral service was conducted at this same church.

169 The Downes papers: Joseph Weiner in the preface to his 1955 book thanks R. L. Downes of Birmingham University for his aid "in connection with Sussex iron work and kindred matters of which he is making a special study" (vii). Downes (died 1981) is not a name familiar in Piltdown circles, and since the present chapter also makes use of his re-

search, some explanation is proper. At the exposure in 1953, Downes, age thirty, was a graduate student on the faculty of Birmingham University. In his dissertation treating the history of England's iron industry, he cited the Beauport statuette as proving that ancient Rome had produced cast iron. When in May 1954, alerted by the fraud publicity, he became aware of Dawson's link to the statuette, he began researching Dawson's antiquarian career. Turning up strong indications of Dawson's earlier impostures, he informed Weiner, then at work on his own book, supplying him a lengthy written account of his findings, styled an appendix. Weiner used only a small portion of this, however, and Downes then began preparing a book-length manuscript of his own on Dawson's early career. This was declined by half a dozen London publishers (the unfinished manuscript tells why: rather inept presentation in both writing and organization of the extensive materials). Downes then dropped his interest in Dawson and Piltdown. Eventually his materials reached the Sussex Archaeological Society, where they became available for study in the summer of 1993. (For bringing the Downes papers to my attention I am indebted to Joyce Crow and Susan Bain of the Society Library, Barbican House.) Unless otherwise indicated, the use I make of Downes' Dawson research is my own.

170 The Beauport statuette: briefly mentioned in Weiner, *Forgery*, 182, from information supplied by R. L. Downes. My treatment is based on facts in the unpublished Downes appendix, A1–10, and other sources as cited.

170 "in Europe at least"—*SAC*, V. 46 (1903), 5.

171 "Much rusted and"—*Proceedings of the Society of Antiquaries of London*, V. 14, 359. On the same page occurs the reference to "human bones." (Downes in his appendix denies that bones were found at Beauport but gives no evidence.) The bones may only have been another trick of Dawson's.

171 "wrought, malleable"—Ibid., 360. Same for the next three quotations.

173 "Dr. Kelner, Analyst"—*SAC*, V. 46 (1903), 9.

173 "If we may speculate"—quoted in Straker, *Weald*, 336, from *SAC*, V. 46.

174 "the most interesting"—Straker, *Weald*, 336.

174 "Notwithstanding Mr."—Ibid., 337.

174 "On Sunday I managed"—Dawson to Crake, September 13, 1893, from a copy in the Downes papers, original at the Hastings Museum library.

175 "the molten iron"—Rock, "Cinder," 172. The fact that Rock was mistaken about Beauport producing molten iron is stated with explanatory detail in Downes, appendix, A7–8.

176 "I have in my possession"—Ibid., 172. The coins are discussed in the Downes appendix, A4–5.

177 "Roman iron manufacture"—Ibid., 174. The suggestion that Rock's article provided the inspiration for the Beauport fake is in Downes, appendix, A8. To that source may now be added another, of equal interest and

equally available to Dawson. In Rock's article is a reference to an earlier paper on Sussex ironwork by M. A. Lower, published in *SAC*, 1894 (II, 169–81). Lower's paper does not mention Beauport, but it does mention the "manufacture of iron" in Roman Sussex, adding that its origins were "shrouded in mystery" and that "little has hitherto been known" of the Sussex iron trade. Lower also mentions the finding of human skeletons under the slag heaps at one spot. Even more intriguing for its resemblance to Piltdown's beginnings is Lower's portrait of the first discovery of Roman activity in the area, made by a certain Reverend Edward Turner:

> In the year 1844 Mr. Turner observed, upon a heap of cinders laid ready for use by the side of the London road, a small fragment of pottery, which on examination proved to be Roman. His curiosity having been excited by so unusual a circumstance, Mr. Turner ascertained, on inquiry, that the cinders had been dug upon Old Land Farm, in his own parish of Maresfield, and immediately contiguous to Buxted. He at once visited the spot and found that the workmen engaged in the digging were exposing to view the undoubted remains of a Roman settlement . . .
>
> A few days previously to Mr. Turner's visit the labourers had opened, in the middle of this field, a kind of grave, about twelve feet in depth, at the bottom of which lay a considerable quantity of broken Roman pottery . . . The digging had been carried on many months prior to Mr. Turner's investigations.

With the change of a few words, that passage could almost be Dawson himself describing to the Geological Society in 1912 what he said happened at Barkham Manor.

178   In 1954 the Beauport statuette was examined by the British Cast Iron Association, when it was confirmed as cast iron, but permitting no precise dating. The amount of sulfur present, however, was taken to indicate a furnace fueled by coke, which meant the statuette "could not have been made before the eighteenth century, and not in Sussex at any date" (Downes, appendix, A9).

178   The flint weapon was described by Dawson in *SAC*, V. 39 (1894), 96–8, under the title "Neolithic Flint Weapon in a Wooden Haft." The quotations are all from this article. I have also consulted the Downes book manuscript, 108–16.

179   The Bexhill boat was reported by Dawson in *SAC*, V. 39 (1894), 161–3, under the title "Ancient Boat Found at Bexhill." The quotations are from this article, except for those from the *Southern Weekly News*, January 21, 1888 (quoted in *SAC*, V. 36, 252–3). The paper added that the boat's "fabric is entirely of oak, the side planks being all pegged together with wood." There is no objective proof of an identity between the boat found by Webb and the one Dawson spotted at the shop of Jesse Young, but it would be a heavy strain on coincidence to have two such ancient

vessels found at about the same time buried in the same place on the same stretch of beach.

180　The transitional horseshoe was described by Dawson in *SAC*, V. 46, 23–4. The quotations are all from this article. I have also consulted the Downes appendix, A14–15.

181　Dawson's smaller frauds of the 1890s: for the clock face see *Antiquity*, November 1981, 220; the Sussex Archaeological Society *Newsletter*, No. 22 (1977), 121; *Antiquarian Horology*, V. 10 (1977), 428–30; and the Downes manuscript, 330–1. For the mace see Baines, *Hastings*, 164, where it is described as "a little staff of silver and ivory, in the shape of the normal tipstaff's mace but ending in a short oar." Baines adds the relevant fact that it was "found in 1902 by the late Charles Dawson in a pawnbroker's shop in Canterbury," citing as his source *The Hastings and St. Leonards Observer*, April 2, 1927. For the ax head, the anvil, and the prick-spur I have used the Downes appendix, A14–15.

181　Dawson's bronze objects from Hastings Castle are described in the Downes appendix, A16, where the *Southern Weekly News* account (November 15, 1890) is quoted from *SAC*, V. 38 (1892), 226.

182　"Mrs. Dawson was rather"—Weiner interview of Lady Woodward, February 1954, Piltdown papers, Natural History Museum. The portion of the interview given in *Papers*, 239, omits this comment.

182　*Plagiaulax dawsoni* is mentioned in Weiner, *Forgery*, 82, 84. Woodward originally described the tooth and the circumstances of its discovery in *Proceedings of the Zoological Society*, V. 24 (1891), 585–6. Downes' manuscript, 76–81, also discusses it, offering a reasonable suggestion about the source of the doctored tooth: Dawson might have procured it from the collection of his sometime mentor in geology, Samuel Beckles of Hastings. Beckles died in 1890, and *Plagiaulax* was produced by Dawson in 1891, after he had helped arrange and catalog the Beckles collection for transmittal to the Natural History Museum. A true fossil tooth resembling that of *Plagiaulax* may well have been in the Beckles collection, which only required some deft filing of the crown to make it a pioneer discovery. Beckles did in fact have a close link to *Plagiaulax*, may even have been the finder of its pioneer fossil as early as 1857, when he was an amateur fossil hunter prospecting in part on behalf of Lyell (see *Sir Charles Lyell's Scientific Journals on the Species Question*, Yale University Press, 1970, 217).

183　"It only seems to"—Woodward, "Tooth," 586. Same for the next quotation. Woodward says that the tooth when found was part of "an irregular mass of communited fish-and-reptile bones, with scales and teeth." But he fails to state whether Dawson brought the entire matrix to the museum or only registered a claim about it. Woodward was surprised that the tooth was "much abraded, evidently the result of wear during the life of the animal." He also finds "remarkable" the tooth's being in some respects so similar to other fossil teeth, leading him to conclude that Dawson's tooth is from "a species larger than any hitherto described." But none of this—so like his action with Piltdown!—was sufficient to slow his rush in naming a new species.

184 "our views do not"—*Geological Magazine*, V. 5, October 1898.

185 "The writer made the"—Dawson, "Dene," 295. Interestingly, and revealingly, Dawson does not say exactly where his descent took place, only that the pit was "in the centre of East Sussex in a very old world neighborhood many miles away from any railway station." His avoidance of names of course was deliberate, strengthening the supposition that he never set foot in a dene hole. In this same article he mentions the Uckfield chemist Samuel Woodhead, who would play a small part in the Piltdown story (he quotes Woodhead on the chemical nature of chalk).

185 "The greater part has"—Weiner, *Forgery*, 182.

185 Downes' study of Dawson's iron articles and related matters is in his appendix, B1–10 and his manuscript, 229–46. His discussion of Dawson's article, "Old Sussex Glass" (*Antiquary*, January 1905), is in his book manuscript, 247–55. Though it is Dawson's first venture into the field of glassworking, in this article he writes with all the authority and apparent personal knowledge of an expert. It confirms again his rare ability to simulate learning: see above, 193–94.

186 "Mr. Baines declares"—Weiner, *Forgery*, 176. Weiner adds: "Mr. R.L. Downes has made a detailed study of the book and is in complete agreement with Mr. Baines."

186 "involved a considerable"—Downes, appendix, B25. Downes' detailed study of Dawson's *Hastings Castle* occupies twenty-five oversize pages (typed) of the appendix. In particular he evaluates Dawson's debt to the Herbert manuscript, summing up: "The Preface implies that the Herbert manuscript consisted of a description of the excavations at the Castle in 1824. The 54 references in the 54 pages of Part V, 9% of the book, appear to confirm this. But apart from the conclusive evidence in Mr. Baines' possession, we have deduced our own reasons for supposing that Herbert wrote [supplied] much of Parts II, III, and IV. Yet in this 76% of the book, Herbert's name is mentioned only 5 times. And not one of these references really suggests that Dawson was indebted to him for more than a sentence or two" (B21). Among smaller anomalies, Downes points out such things as "scores of irrelevant footnotes" and the fact that the book's index, after Chapter Two, goes badly awry, failing to list many names and overlooking many citations of names listed. There is also the fact that while the book's title page bears a date of 1909, publication was actually in mid-1910. Most likely this means that there was a last-minute postponement of plans, probably concerning something about the book's text, some urgently needed alteration or addition.

187 Dawson on the Bayeux tapestry: *Antiquary*, V. 43 (January 1907), 253–8, and August 1907, 288–92. Downes suggests that this paper was a by-

product of Dawson's work on Hastings Castle, and also notes that it is cited in *The Bayeux Tapestry* by Eric Maclagan, Penguin, 1945.

187 Dawson on the "Red Hills"—*Antiquary*, V. 47 (1911), 128–32. Same for the quotation, 252.

188 Dawson and the Heathfield gas: first mentioned in Weiner, *Forgery*, 178–80, and has often been referred to since, always with some distortion of fact. My account is based on the original materials from *Nature*, *Journal of the Geological Society*, and the Downes book manuscript, 85–96.

188 "I am making experiments"—Dawson, *Nature*, 151. He explains that the pipeline has been capped "about fifteen months." Two photographs accompany the letter.

189 "then the gas must vary"—Dawson, "Gas," 574. This apparently spontaneous remark, evidently showing an easy familiarity with *foreign* natural gas fields, is a good example of Dawson's trick of appearing to have detailed knowledge on all sorts of subjects that might have seemed well beyond his scope. Southwest England now has a few small oil fields.

190 The Pevensey brick: my account of this is based on the original materials cited and on the Downes book manuscript, 256–69.

191 The Salzman tile from Pevensey was reported by him in *SAC*, V. 51 (1908), 112. Further Pevensey excavations were reported in *Antiquary*, April 1908, where it is stated that no more tiles had been found.

191 "I was fortunate enough"—Dawson, *Proceedings*, 411.

192 The 1972 thermoluminescence tests: reported in Peacock, "Forged," 138–40. "The lettering itself is curious," adds the report. "The rather spidery style is difficult to parallel among the general run of Roman military or civil tile stamps from Britain." Peacock concludes by suggesting that it is time "for a full investigation of Dawson's numerous and often bizarre discoveries," in particular the Beauport statuette and *Plagiaulax*. But like so many others, with that he let the matter drop.

193 "It does not so much"—*Papers*, 123.

193 "I reminded him that"—*Papers*, 94.

193 "The unicorn became"—Dawson to Mr. Parkin, November 4, 1907, from the original at the Public Record Office, Lewes.

194 "Mr. Dawson observed as"—*QJGS*, V. 69 (1913), 116. The occasion (November 6, 1912) was the presentation to the Geological Society of a collection of Wealden fossil plants made by Teilhard and his friend Fr. Pelletier during 1909–11. Confusion, however, is present even here—if confusion is the word. Making the presentation was the respected A. C. Seward, who had studied the specimens for a year, and who stated at the meeting that the small but important collection had been made by Dawson, "with the able assistance" of the two priests. In the report almost all the nearly thirty specimens are listed as from the "Dawson Coll." and Seward also announced that the entire collection would be donated to the Natural History Museum "at Mr. Dawson's wish." But in the discussion that followed, Dawson himself credits the collection to the two

young prelates, who had "devoted nearly the whole of their spare time" to gathering fossil plants and had "displayed an immense amount of industry and perception." How it came about that the innocent Seward gave the lead position to Dawson, with Teilhard and Pelletier trotting along as his assistants, would be a curious thing to know. For some detail in the matter, see *Papers*, 7, 26, 29, 33.

194 "Have you heard of a"—Dawson to Crake, October 23, 1900, from the original at the Hastings Museum.

194 The Eastbourne skeletons: reported in *SAC*, V. 52 (1909), 189–92.

195 The sea serpent: a passing reference to the incident is in Weiner, *Forgery*, 181, but nothing further was known until the original letter was printed in *Quicksilver Messenger* (Brighton), No. 4, Summer 1981.

197 "if they were genuine"—Weiner, *Forgery*, 155.

198 Dawson's luck: Much has been written on the topic of why and how the Piltdown fraud succeeded so well, with emphasis on the obvious failure of the scientists themselves when faced with personal choices. To this subject the overlooked article by Michael Hammond (*Anthropology*, 1979) discussing the scientific background at the time makes an interesting contribution, and also, by implication, helps clarify the true extent and quality of Dawson's luck. As Hammond shows in detail, the then prevalent view of paleontological theory about human evolution—always subtly changing—by 1912 had reached precisely the stage of scientific expectation most welcoming to Piltdown Man, several disparate factors converging. This was Dawson's greatest piece of sheer, unanticipated good fortune. Even with all his careful study and preparation, he could not have been aware of all the scholarly facets involved, could not have fully appreciated the wonderful fit his creation would make in the theoretical picture. For some discussion see Hammond, 55–6, and Lewin, 69–75.

198 Dawson and the arrowhead: *Papers*, 14. For Dawson's letter mentioning the incident see above, 113–14. The date of the arrowhead find at Crowborough was November 30, 1911, when Piltdown was just on the verge of commencing. I am convinced that Dawson meant the arrowhead to quicken Doyle's interest in excavation, preparing him for eager involvement in the fraud. Imagine if he had succeeded in enticing the famous author to dig at Piltdown, perhaps arranging for him to "find" a bit of the skull! This contention should be weighed in light of my final chapter.

Added Note to Chapter Eleven

At different times Dawson was involved in a number of other questionable matters which appear to have been concerned with faked artifacts or doctoring of one kind or another. Two of these may be described.

The Lavant Caves (1893). Subterranean chalk caves in East Lavant were excavated under the supervision of Dawson and a friend, the various small finds representing a peculiar agglomeration of artifacts. There were many items in bronze (buckles, an oil lamp, pins, a needle, ornamental pendants), as well as Samian ware and other pottery, worked flints, human teeth, lead seals, and a small terra-cotta lamp. There was, and still is, much disagreement over the interpretation of the finds and the nature of the caves, though nothing has been done about investigating for possible fraud. Gradually, after being exposed, the caves fell in and were lost to further study, a fact that was criticized as late as 1916. In that year an article in *SAC* (Allcroft) describing the site stated, "It is much to be regretted that timely steps were not taken to underprop at least a small part of the whole . . . the skill of a north country miner would have dealt easily with the matter at the outset, and enabled the whole area to be cleared, searched, and planned" (V. 58, 74). Dawson's own official report of the work was given in a talk to the Sussex Archaeological Society in 1893 (reported in *Sussex Daily News*, August 11, 1893, and *Antiquary*, V. 28, pp. 22, 160). The talk was never published, but the original manuscript is preserved at the West Sussex Public Records Office, Chichester. For further information on the Lavant Caves see SAS *Newsletter*, No. 33, 234, and No. 34, 244; also Weiner, *Forgery*, 172–3.

The Maresfield Map (1912). Accompanying an article by W. V. Crake in *SAC* (V. 55, 279–83) concerning an ancient forge in Maresfield (near Piltdown) is a hand-drawn map of the area that bears the notation "Made by C. Dawson, F.S.A." The caption identifies the map as a "copy" of a map made originally in 1724. Nothing particular was noted about the map at the time, and it was later reprinted at least once (Straker, *Wealden*, 401), but according to good evidence made public in 1974 the map is "wholly fictitious, and of no value for anything depicted at any period" (Andrews, "Purported," 165). Study reveals pseudo-archaic spellings, anachronistic symbols and nomenclature, anomalous topographical features not found in authentic maps of the time, incorrect limits of river navigation, etc. For further discussion see SAS *Newsletter*, No. 15, 67, and *Antiquity*, November 1980, 187; November 1981, 220–2.

CHAPTER TWELVE:
Return to the Pit

Since this chapter is essentially a rehandling, from a different angle, of material covered in previous chapters, citations are given only for the more significant items. A comparison with the same incidents as depicted in chapters two and three will show from what events and circumstances the explanations are derived. For the order, number, and finder of all the Piltdown specimens—human, animal, and flints—see the tables in *Papers*, 41, 98, 118. The table for 1912 assigns two skull fragments to 1908 and 1911. These were claimed by Dawson as turning up in those years, but of course they were never at any time in the ground anywhere.

201 "after much inspection"—Woodward, *Earliest*, 10.

203 "I struck part of"—Dawson, *Naturalist* (1913), 77. Woodward's presence at the critical moment is attested by his saying that *"we both saw half of the human lower jaw fly out"* (*Earliest*, 11, italics mine). Same for "untouched remnants."

205 Samuel Woodhead: By 1908 Dawson had known Woodhead for more than a dozen years and had used his chemical knowledge on at least two prior occasions (the Heathfield gas and the article on dene holes). Obviously he found the chemist to be sufficiently trusting and pliable, so he had no qualms about taking him to the Barkham Manor pit. Woodhead's testing of the cranial fragment for its organic content (see above, 103–4), but never suggesting that the same be done for the jaw, tells how right Dawson was in his evaluation. It is no use to say that Woodhead might have made such a suggestion in private, only to have it negated. Woodhead lived until 1943, and never called publicly for the one test that he, as a chemist, should have known was indispensable.

205 "in order to stir me"—Teilhard, *Hastings*, 191. The fact of Dawson's deliberate snaring of Teilhard for involvement at Piltdown was unwittingly preserved by Woodward in his book, *The Earliest Englishman*. Informed of the young priest's exploring for bones at the Hastings quarry with a friend in 1909, Dawson told the quarry foreman to let the two roam unhindered, and he himself would provide the usual small payments or tips: "At the same time he asked about the customary days and hours of the Frenchmen's visits, and soon made an opportunity to meet them in the quarries" (Woodward, *Earliest*, 9). The meeting of Dawson and Teilhard was not at all the accidental encounter usually portrayed.

206 "In 1913 my staying"—*Papers*, 251. It was during this 1913 visit to England that Dawson took Teilhard to see the Barcombe Mills site (see above, 137), coinciding with Dawson's report of a skull fragment "found" there that July (*Papers*, 70). I suggest that on this visit Dawson actually planted something at Barcombe for Teilhard to "discover," but that the priest's searching eyes simply missed it, and the effort was not repeated because Dawson became occupied with producing the canine at Piltdown. Teilhard's failure at Barcombe, I also feel, was what prompted Dawson to use a grid system ("mapped out in squares") for the finding of the canine. See two notes down.

207 Hargreaves' absence on August 30: no reference in any source, direct or otherwise, places Hargreaves on the scene that day, and Woodward clearly states (*Earliest*, 11) that the actual digging was done by the three searchers themselves. The shovel was mostly wielded by the younger Teilhard, who after laboring for a time in the hot trench "seemed a little exhausted." (For the whole passage, see above, 42–43.) Hargreaves *was* employed at the pit, however, just before August 30 (*Papers*, 76) and just after (*Papers*, 87).

207   "mapped out in squares"—*QJGS*, V. 70, 85. Nowhere in the existing Piltdown literature, or in any published source that I have seen, is there the least hint that a grid system was employed at the excavations before August 30, 1913, the day that Teilhard picked up the canine. If it was in operation before this, why does Dawson mention it only here, and Woodward nowhere? Grid systems in archeological digs today, of course, are meant primarily for plotting the location and relative position of finds. It is clear that this was not the case at Piltdown.

207   Finding of the canine: for Woodward's part see *Earliest*, 11–12. For Teilhard's part see *Paris*, 104–5.

208   "so inconspicuous amidst"—*Papers*, 212.

208   "unlike anything hitherto"—Woodward, *Earliest*, 47.

208   "insufficient to date"—Woodward, "Early Man," 3. Same for the next quotation.

209   "about a foot below"—*QJGS*, V. 71, 144.

209   "beneath part of the"—Woodward, *Earliest*, 12.

209   Walking to Piltdown from Uckfield: in the 1904 *Booker's Guide to Uckfield and District*, in a section on "Walks," the roundtrip distance is given as "about five miles." The route is described as going along "Church street, and the Rocks, and the residence of R. J. Streatfeild, Esq., will be passed on the left. At the end of a sandstone wall will be seen a stile, on the right hand side. This is the commencement of a very pleasant walk through wood and field, terminating on Piltdown Common." An alternate return went by way of Shortbridge.

209   "thrown there by the"—*QJGS*, V. 71, 144. To the list of Woodward's procedural shortcomings in the excavations may be added his total neglect of the diggers who ordinarily worked the pit for road metal, most glaring in the case of the bone implement. This instrument, Woodward felt sure, had been dug up and then discarded by the diggers: "thrown there by the workmen with the other useless debris" (*QJGS*, V. 71, 144). Yet he did not make the least move to gather the testimony of these workmen, even though he regarded the bone implement as of unique importance for dating the entire Piltdown remains. In his subsequent writings on Piltdown he quietly *assumed* discovery of the implement at the pit's bottom by the workmen, producing such wildly misleading statements as "It was found in the yellow mud at the base of the gravel near the human skull" (Woodward, *Guide*, 12).

213   "asked the labourers to"—Woodward, *Earliest*, 6–7. Just before Woodward wrote (1914) his description of the smashed coconut, Dawson wrote and published a completely different version of the same incident (fictional incident, be it remembered). His lengthy article of August 1913 in the *Hastings and East Sussex Naturalist* has the workmen *blindly* breaking and discarding the skull: "Altogether we found nine separate fragments of the skull, not including the mandible, and there can be lit-

tle doubt that when the workmen first dug up the skull it was complete in most of its details, and that it was shattered and mixed with the gravel before any part of it was noticed by them" (77). Weiner, too, caught the contradiction, but could only comment, "The origin of the coconut story sinks into obscurity" (*Forgery*, 129). Even so I am inclined to think that Dawson knew very well what he was doing in fostering two versions of the story. If challenged, the discrepancies between them could be made to appear as the usual confusion to be expected in oral testimony. No doubt Dawson would have explained to any inquirer—in that honest, open way of his!—that the idea of a coconut had dawned on one of the workmen after the skull had been accidentally smashed and discarded by another, one piece only remaining. Something like this he *must* have told to Woodward to enable him to write what he did.

214   Mabel Kenward: her testimony (invalidated on its face, of course, once all the implications of the fraud are recognized) is available in Weiner, *Forgery*, 192; *Papers*, 246–7; Spencer, *Forgery*, 239; Costello, "Beyond," 169; in her own later article, "Red-Letter Days at Piltdown" (1955); and in a letter to the *Daily Telegraph*, February 23, 1955. Aside from slight discrepancies in all these sources, there is the fact that Dawson, in asking permission of the lord of the manor to dig in 1912, actually lied as to his purpose, saying he was after flints (*Papers*, 49–52). In that case he must also have lied to the Barkham Manor tenant, Robert Kenward, Mabel's father. But this he could hardly have done if, as Mabel asserted, her father had been directly concerned in 1908 with the finding and smashing of an intact skull, the supposed "coconut." This is all quite aside from the fact that in 1908, contrary to Dawson's story, *nothing* happened at the pit. Mabel's memories, no doubt to her quite real, were the inevitable result of later reading and much rambling reminiscence.

214   Dawson's story of the initial find: probably he did, as he said, first become aware of the existence of an ancient gravel bed at Barkham Manor while presiding over one of its periodic court barons (an assembly of tenants and others having business with the manor's lord). The records of these gatherings still exist, showing that sessions were held in 1899, 1904, 1907, and 1911 (Weiner, *Forgery*, 128). A single volume, the original records are now stored at the public record office, Lewes, but because of water damage are too fragile to be consulted. (I am grateful to county archivist Roger Davey for showing me the damaged volumes, in which I had hoped to find something to help determine dates, circumstances, etc. Repairs are scheduled but will involve a lengthy process.) At a guess I would choose 1907, with the probability that the pit was first opened for the extraction of road metal that same year, or the one previous. Dawson's elaboration of the story, showing him strolling on the grounds, noting flints sunk in the road, asking his hosts where they came from, then visiting the pit and speaking with the diggers, of course is all embroidery (perhaps adapted from the Lower article: see above, 248).

Finished?

215   Interview with Mrs. Hodgson: September 23, 1993. Also present was her daughter, Mrs. Ruth Niblett, to whom I am grateful for several courtesies.

215   Woodward as suspect: Some early discussion of this possibility, rambling and suppositious, is mentioned in Spencer, *Forgery*, 232. The first open accusation came only in 1994 when Gerrell Drawhorn of the University of California gave a short paper at the annual meeting of the American Association of Physical Anthropologists. The charge, based on the old assumption that Dawson would have needed expert help, is that Woodward supplied all the doctored specimens as well as technical advice. No direct evidence is offered, only various chemical analyses which attempt—again, relying on nothing more than probability—to link the Piltdown specimens with sources available to Woodward (*American Journal of Physical Anthropology, Supplement 18* [1994], 82, an abstract).

216   "Dawson would not tell"—*Papers*, 163, in Woodward to Hrdlicka, October 26, 1926. He adds the phrase already quoted, "I can only infer from other information that I have." Woodward's frank admission that he could not identify the site of Piltdown Two, Dawson having deliberately kept the information from him, was made only in private. But it is supported by a similar statement of his wife's (partially quoted above, 59), which goes even further, actually questioning Dawson's reliability. At her 1954 interview with Weiner, "she was firm that Dawson would not give details of the exact spot, that her husband was most anxious about it, and that D's illness made his inquiries fruitless, and that he spent much time searching for site II specifically." She even thought that her husband, then dead ten years, had begun to doubt that site Two was real at all: perhaps it only "existed in Dawson's imagination" (*Papers*, 239). Of course if her husband really did entertain, even if momentarily, thoughts about the unreality of a site for Piltdown Two, it was illegitimate for him to have proceeded as he did. Further, her putting the blame on Dawson's physical condition was far from accurate. Between his showing Woodward the remains of Piltdown Two and the first indications of his illness, at least six months intervened. Up to May 1916, if not beyond, Dawson remained clear-headed and approachable.

217   The Gregory article: see above, 50, 146.

217   "made a coherent and"—Weiner, et al., "Further," 230. Same for the next quotation. Weiner does concede, however, that Woodward's insistence on the "human features" of the Piltdown jaw and teeth was a very questionable position to take. Present studies, Weiner declares, all emphasize the "astonishing similarity of the mandible to that of a modern orang or chimpanzee" (232), a fact that should at least have slowed Woodward in his rush to decide the question. For an instance of his loose reasoning about the jaw and its dentition, see *Earliest*, 71–2, where

his conclusion hinges on such equivocal phrases as "seems to" and "probably" and "not much doubt."

218 Woodward's delayed presentation of the Piltdown Two specimens to the Geological Society, made in 1917, downplayed the event, making it subsidiary to other things. His paper is entitled "Fourth Note on the Piltdown Gravel, with Evidence of a Second Skull of *Eoanthropus Dawsoni*." The gravels, already much discussed, certainly did not deserve precedence over Piltdown Two (and why "evidence of," where the first paper boldly announced the "Discovery of a Palaeolithic Human Skull"?). In the presentation itself the skewed order is continued. Only after speaking at length about the gravels and some new flint finds, does Woodward arrive at his announcement of Piltdown Two (see above, 59). Openly he states that the new finds were made by Dawson "in the winter of 1914–15," yet he gives no reason why they were held back for two years, which must have left many in his audience at least mildly unsatisfied.

218 On the topic of Woodward's casual attitude, a later statement of his made in the museum's *Guide to the Fossil Remains of Man* (1922), tells its own subtle tale. He writes that the canine was found "lying in the gravel close to the spot whence the lower jaw itself had been disinterred" (22). This appears to claim that both jaw and canine were found on the same level, in close association in the same soil. But of course the jaw was turned up in the earth of the pit bottom, while the canine was picked out of a dispersed spoil heap spread atop undug ground. Since the truth was on record elsewhere, including in Woodward's own hand, the misstatement could not have been quite deliberate. Perhaps it was the awkward result of a convoluted thought process that regarded the gravel of the spoil heap as having been shoveled out of the pit "close to" the jaw's location. But notice how this innocent exaggeration did not work against, but very much in favor of Woodward and *Eoanthropus*. Interestingly, a similar type of error, again working very much in favor of *Eoanthropus*, was committed by Dawson: see above, 226–27.

219 "patch of gravel"—Woodward, *Earliest*, 65. The four lines containing this phrase occur at the end of a paragraph defending linkage of the jaw and cranium. They are presented abruptly and without elaboration: "We are confirmed in this belief by Mr. Dawson's discovery of a similar grinding tooth, together with two fragments of a second Piltdown skull, in a patch of gravel about two miles away from the original spot." With that strangely brief and bare statement, Woodward veered permanently away from discussion of Piltdown Two. Thereafter he made fleeting mentions of it only in *Guide*, 25, and "Early Man," 3.

219 That Woodward from time to time indeed felt nagging reservations about the linkage of jaw and cranium was claimed by two of his personal acquaintances, both qualified scientists. He exhibited "streaks of doubt," recalled one, while another thought he seemed uncomfortable when talking of linkage, voicing "cautious opinions" (Weiner, *Forgery*, 138–9,

and *Papers*, 245–6). In that case, it can be stated that he must have had more than a few anxious moments as he asked himself the very question that later troubled Weiner: just how had the two unusual specimens, so divergent, happened to come together in the same pit? His reply to the question each time it insinuated itself would of course have been a firm restatement of his implicit belief in linkage.

Added Note on Conan Doyle
(See Chapter 8, and pp. 237–39)

In *Pacific Discovery* (Journal of the California Academy of Sciences), Spring 1996, the attack on Conan Doyle as perpetrator is renewed. It builds on the case set out by Winslow in 1983, but in fact it is hard to decide whether the article is meant to be taken quite seriously ("The Case of the Missing Link" by Robert Anderson).

The "evidence" again revolves on Doyle's novel, *The Lost World*, and rests mainly on a claimed parallel between the fictional diagram of a cave in the novel and the actual pattern of roadways and footpaths in the Piltdown area. Also depicted as (somehow) revealing is the fact of a golf course being near Piltdown. Doyle was an avid golfer and in his novel he refers to a total of eighteen caves, a fact which, it is charged, deliberately mirrors "the 18 fairways" on a golf course (grotesque as this sounds, I assure readers that I have not distorted the argument). Several other minor points of "evidence" are no more impressive than that. One of them tries lamely to equate a sighting of the moon in the novel with Moon's Farm near Barkham Manor.

The cave/road parallel is itself wildly misapplied. As the author concedes, the fit of diagram and roadways is not precise but only "roughly matches" (!) the actual pattern, and then with a good deal of tugging. Also, in the dense tangle of roads, streets, footpaths, and highways crowding the Piltdown area—all plainly visible on the Ordnance Survey map supplied by the author—a diagram of almost *any* shape might find an approximate fit. Further, the article uses a recent OS map. But on the OS map for 1912, a part of the supposed pattern is absent.

Added Note on Martin Hinton
(See pp. 101–2, 236)

In *Nature* (May 23, 1996), the case against Martin Hinton—first leveled in 1978 but never an accusation of any substance—is renewed. A canvas trunk bearing his initials was found in an attic of the Natural History Museum ("found," it should be noted, more than twenty years ago and under circumstances that are vague at best). It yielded a few "assorted bones" that had been chemically treated, none of them human but not further identified. Supposedly, the chemicals used are the same as those detected in the Piltdown specimens. This fact, says the accuser, Brian Gardiner, a paleontologist at King's College, proves beyond doubt that Hinton was the "sole hoaxer." Further, declares Gardiner, it is clear that Hinton was able to make use of an "incompetent" (!) Charles Dawson as his innocent dupe, leading him skillfully to the various finds.

If there is any basis at all to the chemical comparison (no details are supplied in the article), its true explanation proves both simple and innocuous. The staining work on the trunk bones represents experiments carried out by Hinton in conjunction with the *exposure* of the fraud in 1953–55. It is known that Hinton, once keeper of zoology at the museum, was in close touch with Oakley and the dozen other scientists involved in the exposure. He can even be heard at the time making pointed suggestions about the fraudulent staining (see *Papers*, 227, 232, 243, 244). This suggestion, incidentally, also explains an overlooked question, *why* the Hinton trunk with its contents was still at the museum thirty years after Hinton's retirement, and fifteen years after his death. Obviously, the materials in it belonged to the museum.

But in any case the charge against Hinton as sole hoaxer is promptly and thoroughly invalidated by some specific facts of the Piltdown excavations as set forth above (see pp. 123–25, and Chapter 12). The timing and placement of the numerous individual plantings in and around the pit (more than forty in all), the carefully managed "discovery" of the separate pieces at widely different times, and the continued manipulation of the unsuspecting diggers, *demand* the forger's repeated presence on the scene as digging progressed during more than two years, the three summers in which finds were made. But in this period, 1912–1914, Hinton was *never* at the Piltdown pit, certainly not during excavations.* At that time he was a young, part-time volunteer on the museum's operating staff fervently hoping for a full-time appointment, which he finally gained in 1921.

The motive assigned Hinton as forger must not be ignored. It seems that his purpose in the deception was one of simple revenge, directed at Arthur Woodward. The cause? In 1910 Woodward hired young Hinton to do some cataloguing at the museum, payment to be made on completion of the job. Hinton, needing the money, asked for a weekly wage instead. He was refused (so Gardiner theorizes in the absence of Woodward's written reply). Thereupon, to get even for this insulting treatment the furious Hinton concocted the whole, elaborate hoax, giving himself up to long years of dire risk and secret, painstaking effort. The suggestion is offered quite soberly.

In charity, I pass over the question of how Hinton as hoaxer might be tied into the Sheffield Park and Barcombe Mills fakes, all of which went straight from the workbench to Woodward via the able hands of Charles Dawson. Similarly, I do not press the matter of Dawson's undoubted claim (of course, *only* a claim) as early as *1908*, expressed to Samuel Woodhead who accompanied him to the site, that workmen had given him bones picked up at Piltdown.

Since Hinton has by now become a stock character in the several freewheeling conspiracy theories already proposed (the two most wonderfully convoluted are listed above, note 95, p. 235), it will be interesting to see how future conspiracy buffs are able to work into the mix the curious contents of the canvas trunk.

---

* He may, of course, have been in the large group of geologists from London that visited the site in July 1913. See above, pp. 39–40.

Selected Bibliography

Only those items that are cited in the notes, or which have a direct bearing on the various phases of the text, supplying essential background information, are included here. The large body of scientific Piltdown literature generated in the period before exposure of the fraud (1953) now mostly possesses only a tangential interest, and pertains more properly to the sociology of science. Listings of this pre-1953 material may be consulted in Miller (1916), Blinderman, Spencer, and Tobias: see below. For Piltdown in its character as a fraud, what follows is the most complete listing of sources in English to date.

Abbott, W., "Pre-Historic Man: The Newly Discovered Link." *Hastings and St. Leonards Observer,* February 1, 1913.

Allcroft, A., "Some Earthworks of West Sussex." *Sussex Archaeological Collections,* V. 58 (1916), 65–90.

Andrews, P., "A Fictitious Purported Historical Map." *Sussex Archaeological Collections,* V. 112 (1974), 165–7.

Baines, M., *Historic Hastings.* Parsons, Hastings, 1955.

Berry, T., "The Piltdown Affair," *Teilhard Newsletter,* V. 13 (July 1980), 12.

Blinderman, C., *The Piltdown Inquest,* Prometheus, New York, 1986.

Booher, H., "Science Fraud at Piltdown: The Amateur and the Priest," *Antioch Review,* V. 44 (1986), 389–407.

Boule, M., "La Paléontologie humaine en Angleterre," *L'Anthropologie,* V. 28 (1915), 157–9.

Bowden, M., *Ape-Men: Fact or Fallacy?*, Sovereign, Kent, 1978.

Broad, W., and N. Wade, *Betrayers of the Truth*, Simon & Schuster, New York, 1982.

Carr, J. D., *The Life of Arthur Conan Doyle*, Harper, New York. 1949.

Chamberlain, A., "The Piltdown Forgery," *New Scientist*, V. 40 (1968), 516.

Clark, W. Le Gros, "Exposure of the Piltdown Forgery," *Proceedings of the Royal Institute*, V. 20 (1955), 138–51.

Clermont, N., "On the Piltdown Joker and Accomplice: A French Connection?" (I), *Current Anthropology*, V. 33 (1992), 587.

Cole, S., *Leakey's Luck*, Collins, London, 1975.

Combridge, J., "Charles Dawson and John Lewis," *Antiquity*, V. 55 (1981), 220–2.

Costello, P., "Teilhard and the Piltdown Hoax," *Antiquity*, V. 55 (1981), 58–9.

———, "The Piltdown Hoax Reconsidered," *Antiquity*, V. 59 (1985), 167–71.

———, "The Piltdown Hoax: Beyond the Hewitt Connection," *Antiquity*, V. 60 (1986), 145–7.

Cox, D., *Arthur Conan Doyle*, Ungar, New York, 1985.

Daniel, G., "Piltdown and Prof. Hewitt," *Antiquity*, V. 60 (1986), 59–60.

Dawson, C., "Neolithic Flint Weapon in a Wooden Haft," *Sussex Archaeological Collections*, V. 39 (1894), 96–8.

———, "Ancient Boat Found at Bexhill," *Sussex Archaeological Collections*, V. 39 (1894), 161–3.

———, "Discovery of a Large Supply of Natural Gas," *Nature*, V. 57 (16 Dec. 1897), 150–1.

———, "On the Discovery of Natural Gas in East Sussex," *Quarterly Journal of the Geological Society of London*, V. 54 (1898), 564–74.

———, "Ancient and Modern Dene Holes," *Geological Magazine*, V. 5 (July 1898), 293–302.

———, "Sussex Ironwork and Pottery," *Sussex Archaeological Collections*, V. 46 (1903), 1–62.

———, "Sussex Pottery: A New Classification," *Antiquity*, (February 1903), 47–9.

———, "Old Sussex Glass: Its Origin and Decline," *Antiquary*, January 1905, 8–11.

———, "Note on Some Inscribed Bricks from Pevensey," *Proceedings of the Society of Antiquaries*, V. 21 (1907), 411–13.

———, "The Bayeux Tapestry in the Hands of Restorers," *Antiquary*, V. 43 (1907), 253–8, 288–92.

———, *History of Hastings Castle*, Constable, London, 1909 (1910).

———, "The Red Hills of the Essex Marshes," *Antiquary*, V. 47 (1911), 128–32.

———, "The Piltdown Skull," *Hastings and East Sussex Naturalist*, March 25, 1913, 73–82.

———, "The Piltdown Skull," *Hastings and East Sussex Naturalist*, June 1915, 182–4.

Dawson, C., and A. Woodward, "On the Discovery of a Palaeolithic Human Skull," *Quarterly Journal of the Geological Society of London*, V. 69 (1913), 117–51.

———, "Supplementary Note on the Discovery of a Palaeolithic Human Skull and Mandible at Piltdown (Sussex)," *Quarterly Journal of the Geological Society of London*, V. 70 (1914), 82–93.

———, "On a Bone Implement from Piltdown," *Quarterly Journal of the Geological Society of London*, V. 71 (1915), 144–9.

Dodson, E., "Was Teilhard a Co-Conspirator at Piltdown?" *The Teilhard Review*, V. 16 (1981), 16–21.

Doyle, A. C., *The Coming of the Fairies*, Doran, New York, 1921.

———, *The Case for Spirit Photography*, Doran, New York, 1923.

———, *Memories and Adventures*, Hodder, London, 1926.

———, *The History of Spiritualism*, Doran, New York, 1926.

———, *The Lost World*, in *Complete Works*, Crowborough ed., Doubleday, New York, 1930.

Elliott, D., *The Curious Incident of the Missing Link*, Toronto, Bootmakers of Toronto, Occ. Papers, No. 2, 1988.

Esbroeck, G. Van, *Pleine Lumière sur l'Imposture de Piltdown*, Cedre, Paris, 1972.

Essex, R., "The Piltdown Plot: A Hoax That Grew," *Kent and Sussex Journal*, July–September 1955, 94–5.

Farrar, R., "The Sherborne Bone Controversy," *Antiquity*, V. 53 (1979), 211–16, and V. 55 (1981), 44–6.

Gould, S. J., "Piltdown Revisited," *Natural History*, V. 88 (1979), 86–97; same in *New Scientist*, V. 82, 42–4.

———, "The Piltdown Conspiracy," *Natural History*, V. 89 (1980), 8–28.

———, "Piltdown in Letters," *Natural History*, V. 90 (1981), 12–30.

———, *Hen's Teeth and Horse's Toes*, Norton, New York, 1983.

Gregory, W. K., "The Dawn Man of Piltdown," *American Museum Journal*, V. 14 (1914), 189–200.

Grigson, C., "Missing Links of the Piltdown Fraud," *New Scientist*, V. 125 (1990), 55–8.

———, "Sir Arthur Keith and the Piltdown Forgery," *Times Literary Supplement*, February 1, 1991.

Hall, T., *Sherlock Holmes and His Creator*, St. Martin's Press, New York, 1977.

Halstead, L., "New Light on the Piltdown Hoax," *Nature*, V. 276 (1978), 11–13.

———, "The Piltdown Hoax: *Cui Bono*?" *Nature*, V. 277 (1979), 596.

Hammond, M., "A Framework of Plausibility for an Anthropological Forgery: The Piltdown Case," *Anthropology*, V. 3 (1979), 47–58.

Harrison, G., "Weiner and the Exposure of the Piltdown Forgery," *Antiquity*, V. 57 (1983), 46–8.

Head, J., "The Piltdown Mystery," *New Scientist*, V. 49 (1971), 86.

Heal, V., "Further Light on Charles Dawson," *Antiquity*, V. 54 (1980), 222–5.

Hedges, R., et al., "Radiocarbon Dates from the AMS System," *Archaeometry*, Vol. 31 (1989), 207–34.

Heizer, R., and S. Cook, "Comments on the Piltdown Remains," *American Anthropologist*, V. 56 (1954), 92–4.

Higham, C., *The Adventures of Conan Doyle*, Hamilton, London, 1976.

Holden, E., "The Lavant Caves," *Sussex Archaeological Society Newsletter*, V. 34 (1982), 244.

Hooton, E. A., *Up From the Ape*, Macmillan, New York, 1931.

——, "The Piltdown Affair," *American Anthropologist*, V. 56 (1954), 287–9.

Howells, W., *Mankind in the Making*, Doubleday, New York, 1959.

Hrdlicka, A., "The Most Ancient Skeletal Remains of Man," *Smithsonian Annual Report*, 1914, 491–519.

——, "The Piltdown Jaw," *American Journal of Physical Anthropology*, V. 5 (1922), 337–47, and V. 6 (1923), 195–216.

——, "The Skeletal Remains of Early Man," *Smithsonian Miscellaneous Collections*, V. 83 (1930), 65–90.

Jones, M., ed., *Fake? The Art of Deception*, British Museum, 1990.

Keith, A., "Discovery of a New Type of Fossil Man," *British Medical Journal*, December 21, 1912, 1,719–20, and December 28, 1912, 1,763–4.

——, "Our Most Ancient Relation," *The Sphere*, V. 53 (1913), 811.

——, "Piltdown: The Most Ancient Skull in the World," *The Sphere*, V. 53 (1913), 76.

——, "The Piltdown Skull," *Nature*, V. 92 (1913), 197–9, 292, 345–6.

——, "The Reconstruction of Fossil Human Skulls," *Journal of the Royal Anthropological Institute*, V. 44 (1914), 12–31.

——, "Significance of the Discovery at Piltdown," *Bedrock*, V. 2 (1914), 435–53.

——, *The Antiquity of Man*, Norgate, London, 1915.

——, *New Discoveries Relating to the Antiquity of Man*, Norgate, London, 1931.

——, "A Resurvey of the Anatomical Features of the Piltdown Skull," *Journal of Anatomy*, V. 75 (1938), 155–85, 234–54.

——, *An Autobiography*, Watts, London, 1950.

Kenward, M., "Red-Letter Days at Piltdown," *Sussex County Magazine*, Summer 1955, 332–6.

Krogman, W., "The Planned Planting of Piltdown: Who? Why?" in *Human Evolution: Biosocial Perspectives*, eds., Washburn and McCown, Cummings, CA, 1978.

Langdon, J., "Misinterpreting Piltdown," *Current Anthropology*, V. 32 (1991), 627–31.

Langham, I., "Talgai and Piltdown: Common Context," *Artefact*, V. 3 (1978), 181–224.

——, "The Piltdown Hoax," *Nature*, V. 277 (1979), 170.

————, "Sherlock Holmes, Circumstantial Evidence, and Piltdown Man," *PAN (Physical Anthropology News)*, V. 3 (1984), 1–5.

Lankester, E. R., *Diversions of a Naturalist*, Methuen, London, 1915.

————, "A Remarkable Flint from Piltdown," *Man*, V. 32 (1921), 59–62.

Leakey, L., *By the Evidence (Memoirs)*, Harcourt, New York, 1974.

Lewin, R., *Bones of Contention: Controversies in the Search for Human Origins*, Simon & Schuster, New York, 1987.

Lower, M., "Iron Works of the County of Sussex," *Sussex Archaeological Collections*, V. 2 (1849), 169–81.

Lukas, M., and E. Lukas, *Teilhard: A Biography*, Doubleday, New York, 1977.

————, "Gould and Teilhard's 'Fatal Error,' " *Teilhard Newsletter*, V. 14 (1981), 4–6.

————, "The Haunting," (reply to Gould), *Antiquity*, V. 57 (1983), 7–11.

Lyell, C., *The Antiquity of Man*, Murray, London, 1863.

Lyne, W., "The Significance of the Radiographs of the Piltdown Teeth," *Proceedings of the Royal Society of Medicine*, V. 9 (1916), 33–6.

McCann, T., "Charles Dawson and the Lavant Caves," *Sussex Archaeological Society Newsletter*, V. 33 (1981), 234.

McCulloch, W., "Gould's Piltdown Argument," *Teilhard Perspective*, V. 16 (1983), 4–7.

McCurdy, G., "Significance of the Piltdown Skull," *American Journal of Science*, V. 35 (1913), 315–20.

————, "The Man of Piltdown," *American Anthropologist*, V. 16 (1914), 331–6.

Matthews, L., "Piltdown Man: The Missing Links," *New Scientist*, Vols. 90, 91 (1981), 280–2; 376; 515–16; 578–9; 647–8; 710–11; 861–2; 26–8.

Millar, R., *The Piltdown Men*, Gollancz, London, 1972.

Miller, G. S., "The Jaw of Piltdown Man," *Smithsonian Miscellaneous Collections*, V. 65 (1915), 1–31.

————, "The Piltdown Jaw," *American Journal of Physical Anthropology*, V. 1 (1918), 25–52.

————, "The Piltdown Problem," *American Journal of Physical Anthropology*, V. 3 (1920), 585–6.

Moir, J., "The Piltdown Skull," *Antiquary*, January 1914, 21–3.

Nordon, P., *Conan Doyle*, Murray, London, 1966.

Oakley, K., "Fluorine Tests on the Piltdown Skull," *Proc. Geol. Soc. London*, (December 14, 1949, 1950), 29–31.

————, "Artificial Thickening of Bone and the Piltdown Skull," *Nature*, V. 187 (1960), 174.

————, "The Piltdown Skull," *New Scientist*, V. 40 (1969), 154.

————, "The Piltdown Problem Reconsidered, *Antiquity*, V. 50 (1976), 9–13.

————, "The Piltdown Stains," *Nature*, V. 278 (1979), 302.

————, "Piltdown Man," *New Scientist*, V. 92 (1981), 457–8.

Oakley, K., and H. de Vries, "Radiocarbon Dating of the Piltdown Skull and Jaw," *Nature*, V. 184 (1959), 224–6.

Oakley, K., and C. Groves, "Piltdown Man: Realization of Fraudulence," *Nature*, V. 169 (1970), 789.

Oakley, K., and C. Hoskins, "New Evidence on the Antiquity of Piltdown Man," *Nature*, V. 165 (1950), 379–82.

Osborn, H. F., *Men of the Old Stone Age*, Scribner, New York, 1915.

———, "The Dawn Man of Piltdown," *Natural History*, V. 21 (1921), 577–90.

———, *Man Rises to Parnassus*, Princeton, NJ: 1928.

Osborn, H. F., and W. Gregory, "The Dawn Man," *McClure's*, V. 55 (1923), 19–28.

Payne, M., *Crowborough: The Growth of a Wealden Town*, Brewin, Sussex, 1983.

Peacock, D., "Forged Brick Stamps from Pevensey," *Antiquity*, V. 47 (1978), 138–40.

Postlethwaite, F., "Piltdown Man," (letter), London *Times*, 25 Nov. 1953.

Pycraft, W., "The Jaw of the Piltdown Man," *Scientific Progress*, V. 11 (1917), 389–409.

Ray, J. E., "Skeleton Found Near Eastbourne," *Sussex Archaeological Collections*, V. 52 (1909), 189–92.

Rock, J., "Ancient Cinder-Heaps in East Sussex," *Sussex Archaeological Collections*, V. 29 (1879), 167–80.

Rodin, A., and J. Key, *Medical Casebook of Dr. Arthur Conan Doyle*, Kreiger, FL, 1984.

Salzman, L., "Excavations at Pevensey," *Sussex Archaeological Collections*, V. 51 (1908), 99–114.

Schmitz-Moormann, K., "The Gould Hoax and the Piltdown Conspiracy," *The Teilhard Review*, V. 16 (1981), 7–14.

Shipman, P., "On the Trail of the Piltdown Fraudsters," *New Scientist*, V. 128 (1990), 52–4.

Smith, G. E., "The Piltdown Skull," *Nature*, V. 92 (1913), 131; 267–8; 318–19; 468–9.

———, "On the Exact Determination of the Median Plane of the Piltdown Skull," *Quarterly Journal of the Geological Society of London*, V. 70 (1914), 93–4.

———, "Prehistoric Man," *Proceedings of the Royal Philosophical Society of Glasgow*, V. 45 (1914), 17–27.

———, "Primitive Man," *Proc. British Acad.*, V. 7 (1917), 1–50.

———, *The Evolution of Man*, Humphry, London, 1924.

———, "The Reconstruction of the Piltdown Skull," *Proc. Anat. Soc.*, V. 59 (1925), 38–40.

———, "Human Palaeontology," (review of Keith book), *Nature*, V. 127 (1931), 963–7.

———, *The Search for Man's Ancestors*, Watts, London, 1931.

Sollas, W. J., *Ancient Hunters and Their Modern Representatives*, Macmillan, London, 1915.

Speaight, R., *Teilhard de Chardin*, Collins, London, 1967.

Spencer, F., ed., *The Piltdown Papers*, Oxford, London, 1990.

Spencer, F., *Piltdown: A Scientific Forgery*, Oxford, London, 1990.

Straker, E., *Wealden Iron*, Bell, London, 1931.

Teilhard de Chardin, P., "*Le Cas de l'homme de Piltdown*," *Rev. des Questiones Scientifiques*, V. 77 (1920), 149–55.

———, *Fossil Men: Recent Discoveries and Present Problems* (pamphlet), Vetch, Peking, 1943.

———, *Letters From Paris*, Herder, New York, 1967.

———, "Cosmic Life," in *Writings in Time of War*, Harper, New York, 1968.

———, *Letters From Hastings*, Herder, New York, 1968.

———, *The Heart of the Matter*, Harcourt, New York, 1979.

Thackeray, J., "On the Piltdown Joker and Accomplice: A French Connection?" (II). *Current Anthropology*, V. 33 (1992), 587–9.

Thomson, K., "Piltdown Man—The Great English Mystery Story," *American Scientist*, V. 79 (1991), 194–201.

Tobias, P., "An Appraisal of the Case Against Sir Arthur Keith," *Current Anthropology*, v. 33, no. 3 (June 1992), 243–93.

Underwood, A., "The Piltdown Skull," *British Dental Journal*, V. 56 (1913), 650–2.

Vere, F., *The Piltdown Fantasy*, Cassell, London, 1955.

Washburn, S., "The Piltdown Hoax," *American Anthropologist*, V. 55 (1954), 259–62.

———, "The Piltdown Hoax: Piltdown 2," *Science*, V. 203 (1979), 955–7.

Waterston, D., "The Piltdown Mandible," *Nature*, V. 92 (1913), 319.

Weiner, J. S., *The Piltdown Forgery*, Oxford, London, 1955.

———, "Grafton Elliot Smith and Piltdown," *Symposium of the Zoological Society of London*, V. 33 (1973), 23–6.

———, "Piltdown Hoax: New Light," *Nature*, V. 277 (1979), 10.

———, "Teilhard and Piltdown," in *Teilhard and the Unity of Knowledge* by King, M., and J. Salmon, Paulist Press, New York, 1983.

Weiner, J. S., et al., *The Solution of the Piltdown Problem*, Bulletin of the British Museum (Natural History), London, 1953.

———, *Further Contributions to the Solution of the Piltdown Problem*, Bulletin of the British Museum (Natural History), London, 1955.

Weiner, J. S., and J. Emerson, "The Piltdown Mystery," London *Times*, January 9, 16, 23, 1955.

Weiner, J. S., and K. Oakley, "The Piltdown Fraud," *American Journal of Physical Anthropology*, V. 12 (1954), 1–7.

White, H. O., *The Geology of the Country Near Lewes*, Geological Survey of England, 1926.

Winslow, J., and A. Meyer, "The Perpetrator at Piltdown," *Science 83*, September 1983, 33–43.

Woodward, A. S., "On a Mammalian Tooth from the Wealden Formation at Hastings," *Proceedings of the Zoological Society of London*, V. 24 (1891), 585–6.

————, "On Some Mammalian Teeth from the Wealden of Hastings," *Quarterly Journal of the Geological Society of London*, V. 67 (1911), 278–81.

————, "Note on the Piltdown Man," *Geological Magazine*, V. 10 (1913), 433–4.

————, "On an Apparently Palaeolithic Engraving on a Bone from Sherborne (Dorset)," *Quarterly Journal of the Geological Society of London*, V. 70 (1914), 100–103.

————, *A Guide to the Fossil Remains of Man in the British Museum (Natural History)*, British Museum, London, 1915.

————, "Charles Dawson," (obituary), *Geological Magazine*, V. 3 (1916), 477–9.

————, "Early Man," *Geological Magazine*, V. 4 (1917), 1–4.

————, "Fourth Note on the Piltdown Gravel, With Evidence of a Second Skull of *Eoanthropus Dawsoni*," *Quarterly Journal of the Geological Society of London*, V. 73 (1917), 1–10.

————, "The Second Piltdown Skull," *Nature*, V. 131 (1933), 86.

————, "Recent Progress in the Study of Ancient Man," *Report of the British Association for the Advancement of Science*, 1935, 129–42.

————, "Geographical Distribution of Early Man," *Geological Magazine*, V. 81 (1944), 49–57.

————, *The Earliest Englishman*, Watts, London, 1948.

Wright, R., "Samuel Allison Woodhead," (obituary), *The Analyst*, V. 68 (1943), 297.

Index

ABOUT THE AUTHOR

JOHN EVANGELIST WALSH has written a dozen books of history and biography based on original research. In addition to such topics as the invention of the airplane by the Wright brothers and the search for St. Peter's body in Rome, he has probed the lives and careers of such figures as Robert Frost, John Paul Jones, and Emily Dickinson. His book *Poe the Detective* was awarded an Edgar by the Mystery Writers of America. In 1994 his investigation of the legendary romance between Abraham Lincoln and Ann Rutledge, *The Shadows Rise*, was a finalist for the annual Lincoln Prize of Gettysburg College. Two of his books, *Night on Fire* and *One Day at Kitty Hawk*, were condensed in *Reader's Digest*.

Mr. Walsh is married, has four grown children, and lives with his wife, Dorothy, in Monroe, Wisconsin.

ABOUT THE TYPE

The text of this book was set in Janson, a misnamed typeface designed in about 1690 by Nicholas Kis, a Hungarian in Amsterdam. In 1919 the matrices became the property of the Stempel Foundry in Frankfurt. It is an old-style book face of excellent clarity and sharpness. Janson serifs are concave and splayed; the contrast between thick and thin strokes is marked.